石油高职高专规划教材

English for City Gas
城市燃气专业英语

王立柱　杨秀华　主编

石油工业出版社

内容提要

本书用英文对城市燃气进行了全方位的介绍,通过八个模块十九个任务,涵盖了城市燃气基础知识、城市燃气发展概况、城市燃气管网布线、城市燃气管网调峰、城市燃气场站工艺、城市燃气管道敷设、城市燃气场站安全管理、天然气汽车、LNG 与 CNG 等与城市燃气工程技术专业相关的各个领域,有助于培养学生的英语应用能力。

本书既可满足高职高专城市燃气工程技术专业及油气储运类相关专业学生的专业英语学习需求,也可作为石油天然气行业相关从业人员的自学或培训教材。

图书在版编目(CIP)数据

城市燃气专业英语 / 王立柱, 杨秀华主编. —北京:
石油工业出版社, 2017.8 (2024.2 重印)
石油高职高专规划教材
ISBN 978-7-5183-2051-6

Ⅰ. ①城… Ⅱ. ①王… ②杨… Ⅲ. ①城市燃气－英语－高等职业教育－教材 Ⅳ. ① TU966

中国版本图书馆 CIP 数据核字(2017)第 179823 号

出版发行:石油工业出版社
(北京市朝阳区安外安华里 2 区 1 号楼 100011)
网　　址:www.petropub.com
编辑部:(010)64251362　图书营销中心:(010)64523633
经　销:全国新华书店
排　版:北京点石坊文化发展有限责任公司
印　刷:北京中石油彩色印刷有限责任公司

2017 年 8 月第 1 版　2024 年 2 月第 2 次印刷
787 毫米 ×1092 毫米　开本:1/16　印张:12.5
字数:326 千字

定价:30.00 元
(如发现印装质量问题,我社图书营销中心负责调换)
版权所有,翻印必究

前言 Preface

随着我国城镇燃气产业的快速发展，燃气行业从业人员的需求大量增加，国家和燃气企业都将在燃气工程施工、燃气输配与运营管理、燃气应用、安全供气等方面迎来越来越多的挑战和考验。同时，各相关岗位都需要大量专业技术娴熟又具备较高英语听说水平的人才。

为了适应我国城镇燃气产业发展趋势并满足高等职业技术教育的迫切需要，石油工业出版社联合各石油高职院校，于 2015 年 11 月召开了高职高专油气储运和城市燃气专业"十三五"规划教材研讨会，本教材的编写计划就是在该次会议上确定的。本教材以满足高职高专的城市燃气工程技术专业学生专业英语学习需求为目的，以丰富的专业内容，用英文详细地说明了城市燃气工程技术专业相关各领域的知识，并在各模块加强了对"听、说、读、写、译"能力的培养，有助于提高学生的综合能力。

本书由王立柱、杨秀华任主编，郑志霞任副主编，具体编写分工如下：第一模块由杨秀华（河北石油职业技术学院）编写，第二、第六、第七模块由王立柱（天津石油职业技术学院）编写，第三模块由彭朋（大庆职业学院）编写，第四模块由何平（山东胜利职业学院）编写，第五、第八模块由郑志霞（天津石油职业技术学院）、曹丽丽（渤海石油职业学院）编写。郑殿福（天津石油职业技术学院）任本书主审，王立柱负责了全书的统稿工作。

本书编写中引用或借鉴的文献和资料较多，对编者的帮助极大，在此一并向相关作者表示诚挚的谢意。

由于编者水平有限，错误之处在所难免，恳请诸位读者批评指正。

编 者
2017 年 6 月

特别提示：本书听力部分内容，请扫描对应二维码。授课教师索取授课用听力资源包，请发邮件至 wlzshzh@163.com。

目 录
Contents

Module 1　An Introduction to City Gas

Task 1　Classification of City Gas　　　　1
Task 2　Consumer Demand for City Gas　　　　7

Module 2　City Gas Network

Task 3　Classification of City Gas Network　　　　12
Task 4　Supply of City Gas　　　　17
Task 5　Gas Network Peak Shaving　　　　22

Module 3　Process in Gas Transmission and Distribution

Task 6　Hydraulic Calculation for Gas Network　　　　28
Task 7　Station Processes　　　　37

Module 4　Equipment and Facilities

Task 8　Equipment and Facilities for Stations　　　　41
Task 9　Accessory Equipment for Gas Pipeline　　　　48
Task 10　Application Equipment for Gas users　　　　54

Module 5　Gas Network Construction

Task 11　Gas Pipeline Construction　　　　59
Task 12　Quality Supervision and Inspection for Gas Pipeline　　　　65

Module 6 Operation and Maintenance

    Task 13 Station Management   70
    Task 14 Gas Pipeline Maintenance   75

Module 7 Safety Management

    Task 15 Safety Management for Gas Stations   80
    Task 16 Safety Management for Gas Pipeline   85
    Task 17 Safety Emergency Response   89

Module 8 LNG and CNG

    Task 18 Liquefied Natural Gas   95
    Task 19 Compressed Natural Gas   100

Vocabulary   105
Translations and Answers   121
References   181
Appendix Listening Materials   182

Module 1

An Introduction to City Gas

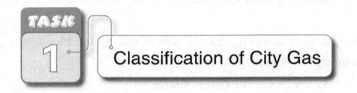

Part 1 Listening

Listen to the passage about the definition and classification of city gas and fill in the blanks with proper words.

City gas or urban gas refers to any type of 1) _____ gas which meets the **normative** gas quality requirements, and is supplied to meet the needs of residential, commercial and industrial users, generally including 2) _____ gas, liquefied petroleum gas (LPG), artificial gas and biogas.

Natural gas is a gaseous fossil fuel found in oil fields, natural gas fields and 3) _____ beds. It is the result of decay of animal remains and plant remains that have occurred over millions of years. The primary 4) _____ of natural gas is **methane**, and it also contains gaseous **hydrocarbons** such as **ethane**, **propane** and **butane**, as well as other non-hydrocarbon 5) _____.

LPG, one of the main sources of city gas, is obtained as a 6) _____ during the process of exploitation of natural gas and petroleum or petroleum refining. The main components of LPG are propane, **propene**, butane and **butene**.

Artificial gas refers to the 7) _____ gases made from some solid or liquid fuel and produced through all kinds of hot working. Based on different raw materials and processing methods, it can be divided into coal gas and 8) _____ gas.

Various organic substances, such as proteins, **cellulose**, 9) _____, **starch**, etc., ferment in the absence of air, and produce a type of combustible gas under the action of microorganisms, which is called marsh gas (biogas), and can be divided into two categories: natural and 10) _____.

Part 2 Speaking

1. Useful sentences

(1) According to different origins and production methods, fuel gas can be divided into four categories: natural gas, artificial gas, liquefied petroleum gas and **biomass** gas.

(2) LPG is not only an important fuel to our residential, commercial, industrial and automobile users, but also an important raw material in chemical production.

(3) Biomass gas, generally known as "biogas", can be divided into two categories: natural and artificial.

(4) Natural biogas such as **sludge** biogas, **sewer** biogas and biogas from mines and coal seams, is naturally formed of organic matters in nature.

(5) Artificial biogas is a renewable energy source.

(6) Calorific value is an important indicator in town gas quality analysis.

(7) Natural gas, liquefied petroleum gas or artificial gas, or even the same type of gas, has different compositions and calorific values owing to different origins. Sometimes there may also be great differences in compositions and calorific values.

(8) According to the calorific value, gas is generally divided into high calorific value gas (HCV gas), medium calorific value gas (MCV gas) and low calorific value gas (LCV gas).

(9) Parameters affecting the combustion characteristics of the gas mainly include **Wobbe Index** and **combustion potential** in the nature of the gas.

(10) Wobbe Index is an indicator to reflect the constant thermal load of gas appliances, which is used to analyze and control the gas interchangeability.

(11) **Combustion potential** is a composite indicator that can determine the gas combustion characteristics in a more comprehensive way, reflecting the combustion steady state of a gas appliance.

2. Pair-work

Talk with your partner about the classification of city gas with the help of the above expressions.

Module 1　An Introduction to City Gas

Part 3　Reading

Basic Properties of City Gas

Fuel gas is a mixture composed of a variety of combustible and non-combustible gases, and its main properties include physical properties, **thermodynamic** properties and combustion characteristics.

Main Physical Properties of Fuel Gas

The main physical properties of fuel gas generally refer to the average relative **molecular** mass, density and relative density (specific gravity), **viscosity**, dew point, boiling point and critical parameters, etc., of the gas.

Thermodynamic properties of fuel gas

The thermodynamic properties of the gas include latent heat of vaporization, specific heat capacity, coefficient of thermal (heat) conductivity, etc. Latent heat of vaporization refers to the heat absorbed when the liquid of a unit mass (1kg) turns into the vapor in **equilibrium** state. Under the condition of no phase change and chemical reaction, the heat required to raise unit mass of a substance by unit temperature interval (1K) is known as the specific heat capacity (or specific heat) of the substance. The coefficient of thermal (heat) conductivity is a characteristic parameter of the ability of a substance to conduct heat, meaning the heat conducted per hour when the temperature per meter decreases by 1K in the direction of thermal conductivity. The thermal conductivity of gas increases with the increase of temperature and pressure.

Combustion Characteristics of Fuel Gas

(1) Calorific value: The quantity of heat produced by the complete combustion of unit volume or mass of a fuel is known as calorific value, usually expressed in joules per kilogram. calorific value is an important indicator for the correct selection of a gas appliance, as well as for the evaluation of gas quality, which is divided into high calorific value and low calorific value.

(2) Theoretical amount of air requirement and **flue** gas: As is known, gas burning needs a right amount of oxygen, not too much or too little. The oxygen required for combustion is generally obtained directly from the air. Theoretical air requirement refers to the amount of air that a standard cubic meter of gas needs for complete combustion according to the combustion reaction equations. The higher the calorific value of the gas is, the more the theoretical air requirement will be. The air requirements are quite different for the combustion of the same volume of liquefied petroleum gas, natural gas and coke oven gas. The air required for LPG combustion is about three times that for natural gas, and six times that for coke oven gas. Flue gas is the product of gas combustion. When the amount of theoretical air requirement is supplied, the amount of smoke produced after complete combustion of gas is called the theoretical amount of flue gas. Theoretical flue gas is composed of carbon dioxide, sulfur dioxide, **nitrogen** and water vapor,

and sometimes carbon monoxide in incomplete combustion.

(3) Ignition temperature: Any combustible gas in contact with oxygen under certain conditions will have an oxidation reaction, and the moment of it starting to catch fire is known as ignition, when the stable oxidation reaction turns into an unstable one. Ignition temperature is the lowest temperature at which the mixture of the combustible gas and air (or oxygen) starts the combustion reaction.

(4) Explosion limit: The range of concentration of the mixture of combustible gas and air (or oxygen) with which ignition and then explosion can occur is known as the explosion limit, and the minimum concentration is called lower limit, while the maximum concentration is called upper limit.

(5) Flame **propagation** speed: **Perpendicular** to the surface of the combustion, the propagation speed at which the flame spreads towards the unburned gas is called the flame propagation speed, also referred to as the combustion speed or burning rate. It is one of the basic parameters to determine the interchangeability of the gas, related to the nature of the gas, the composition of gas-air mixture, temperature and pressure.

(6) Wobbe Index and combustion potential: Wobbe Index was used to determine the interchangeability of fuel gas at the beginning when the issue of interchangeability was put forward. When there are great differences between the displacing gas and displaced gas (reference gas) in terms of the chemical, physical properties and combustion characteristics, the burning rate of the gas will be changed largely. In this case, only using Wobbe Index can not control gas interchangeability, it needs to be controlled with the **combustion potential**, a composite indicator reflecting the combustion steady state of a gas appliance, which reflects the tendency of floating flame, yellow flame, backfire and incomplete combustion produced by the fuel gas.

Questions

(1) Have you known the basic properties of city gas after reading the passage? Can you list some of the physical properties or chemical properties?

(2) Try to find the definitions of the following terms in the passage:

latent heat of vaporization

thermal conductivity

calorific value

theoretical air requirement

theoretical flue gas

ignition temperature

explosion limit

flame **propagation** speed

Part 4   Translating

Read the following passage about the development of urban gas and then translate it into Chinese.

General Situation of City Gas Development

As an important part of urban infrastructure, urban gas not only serves to improve the quality of people's life, the natural and social environment, but also has increasingly become the basic industry in the national economy with a pilot role and an overall importance.

Compared with developed countries, China's urban gas starts relatively late. Generally speaking, the development of modern urban gas industry in China has gone through the following three stages:

First stage: before the 1980s, driven by large-scale development of the country's steel industry and supported by the national energy saving funds, a number of urban gas utilization projects were built up all over the country, and gas pipeline facilities were constructed in many cities and towns. At this stage, the industry mainly focused on the development of coal gas, and the gas users were much more than before and thus the amount of gas supply increased greatly.

Second stage: from the 1980s to the early 1990s, LPG (Liquefied Petroleum Gas) and natural gas have been developed rapidly, forming a pattern of coexistence of a variety of gas supplies. While at the same time, a problem was rising that the existing domestic resources could not meet the needs of urban development and economic construction, resulting in the import of LPG in Guangdong and other economically developed coastal areas where the lack of energy has become a big problem. So far, LPG resources both at home and abroad are more fully utilized, and LPG has become one of the main sources of urban gas.

Third stage: since the late 1990s, with the challenges of China's energy structure, the natural gas supply period started, represented by "Shaan-Gan-Ning natural gas to Beijing". And afterwards, a number of national key projects like Se-Ning-Lan, WEPP and Zhong-Wu gas pipelines have been completed and put into production, providing a **premise** and unprecedented opportunities for urban gas pipeline development, which also marks the era of natural gas of our urban gas. The joint promotion of CNG (compressed natural gas), domestic LNG (liquefied natural gas) and imported LNG has brought new prospects for the development and utilization of multi-sources urban gas.

Part 5   Vocabulary

normative['nɔːmətɪv]	*adj.* 规范的，标准的
methane['miːθeɪn;'meθeɪn]	*n.* 甲烷；沼气

hydrocarbon[ˌhaɪdrə(u)ˈkɑ:b(ə)n]	n. 碳氢化合物
ethane[ˈi:θeɪn;ˈeθ-]	n. 乙烷
propane[ˈprəupeɪn]	n. 丙烷
butane[ˈbju:teɪn]	n. 丁烷
propene[ˈprəupi:n]	n. 丙烯
butene[ˈbju:ti:n]	n. 丁烯
cellulose[ˈseljuləuz;-s]	n. 纤维素；(植物的) 细胞膜质
starch[stɑ:tʃ]	n. 淀粉；刻板，生硬
biomass[ˈbaɪə(u)mæs]	n. 生物量；生物质
sludge[ˈslʌdʒ]	n. 烂泥；泥泞；泥状雪；沉淀物
sewer[ˈsu:ə;ˈsju:ə]	n. 下水道；阴沟
combustion[kəmˈbʌstʃ(ə)n]	n. 燃烧，氧化
thermodynamic[ˌθɜ:məudaɪˈnæmɪk]	adj. 热力学的；使用热动力的
molecular[məˈlekjulə]	adj. 分子的；由分子组成的
viscosity[vɪˈskɒsɪtɪ]	n. 黏性，黏度
equilibrium[ˌi:kwɪˈlɪbrɪəm;ˌekwɪ-]	n. 均衡；平静
flue[flu:]	n. 烟道；暖气管
nitrogen[ˈnaɪtrədʒən]	n. 氮
propagation[ˌprɒpəˈgeɪʃən]	n. 传播；繁殖；增殖
perpendicular[ˌpɜ:p(ə)nˈdɪkjulə]	adj. 垂直的，正交的；直立的 n. 垂线；垂直的位置
premise[ˈpremɪs]	n. 前提
Wobbe Index	沃泊指数；华白数
combustion potential	燃烧势

Module 1 An Introduction to City Gas

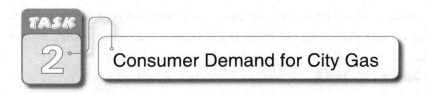

Consumer Demand for City Gas

Part 1 Listening

Listen to the passage about the applications of city gas and fill in the blanks with proper words.

With the development of gas industry, especially the significant exploitation and utilization of natural gas, gas has become an important **pillar** of 1) _____ supply. According to the characteristics of users' using gas, city gas users generally include the following types:

Residential Users

Residential users refer to the 2) _____ gas users who use gas as the fuel for cooking and preparing hot water. They are one of the basic users of city gas supply, and require continuous and 3) _____ gas supply.

Commercial Users

Commercial users are another basic user of city gas supply. They refer to the gas users who use gas for cooking or preparing hot water in commercial facilities or 4) _____ buildings, including staff canteens, **catering** industry, kindergartens, nurseries, hospitals, hotels, barber shops, baths, laundries, supermarkets, offices, research institutes, universities, secondary schools, and primary schools, etc. In schools and research institutes, gas is used in laboratories, as well as for cooking, hot water, and shower.

Industrial Production Users

Industrial production users are those who take gas as the fuel for industrial production. The gas consumption of these users is usually large and 5) _____.

Heating, Ventilation and Air Conditioning Users

They are the users taking gas as the fuel for heating and cooling, and they are 6) _____ load to gas supply, so there must be effective measures to balance the uneven gas requirement in different seasons.

Gas Vehicle Users

These users are those who use gas as the automobile 7) _____. Developing gas vehicles is one of the effective measures to reduce urban air pollution. In addition, gas has an obvious

- 7 -

advantage over petrol in terms of price.

Other Users

Other users mainly include two parts: one is the amount of pipeline 8) _____ due to external damage, natural corrosion, improper use, production **venting** and other factors; the other part is the volume exceeding the original gas volume calculated, owing to new developments unforeseen.

Thermal Power Plants

When power plants use gas as the fuel for **peaking** and generation, urban gas 9) _____ should also include the gas consumption in power stations. Converting clean-burning natural gas to electrical energy with zero discharge of pollutants is a major development direction of natural gas applications.

Data shows that natural gas has begun to be applied and developed in greenhouses 10) _____ flowers and vegetables, grain drying and storage, deep processing of agricultural products, biotechnology, **pharmaceutical**s, **pesticide**s, and gas fuel cell lights, etc., which will lead to continuous expansion and more detailed classification of urban gas users.

Part 2 Speaking

1. Useful sentences

(1) The urban gas in China was mainly used in residents, industry and commerce, while at present it is rapidly expanding into such application fields as gas power generation, gas air conditioners, gas vehicles and chemical-based industry.

(2) The amount of gas consumed by individual residential users is quite small, and the gas usage is not regular.

(3) The amount of gas consumed by commercial users is not very large, and the gas usage is quite regular.

(4) Industrial production users require a large amount of gas regularly.

(5) Heating, ventilation and air-conditioning users have outstanding features of uneven gas usage, while the gas requirement is relatively stable during the heating period.

(6) Gas consumption of gas vehicles users is related to the number of urban gas vehicles and their operation details, not much dependent on the change of external factors like seasons.

(7) Thermal power plants consume a very large amount of gas, which is not included in the gas consumption of the planning city gas pipeline network, and should be considered as an independent project.

2. Pair-work

With the help of the above expressions and what you have learned in listening, make up a dialogue about the application of city gas with your partner.

Part 3 Reading

Quality Requirements for City Gas

Before entering the transmission and distribution network and being supplied to users, the city gas should meet the basic requirements, for instance, relative stable calorific value, low toxicity and few **impurities**, and meet certain quality indicators and remain relatively stable, which has great significance for reducing the pipeline corrosion, pipeline blocking and environment pollution, as well as for the safety of city gas distribution system and gas using, and ensuring the economic **rationality** of the system.

Main Impurities in City Gas

Impurities contained in each type of city gas are not the same, owing to different sources and methods of preparation. Main impurities in artificial gas, for instance, include **tar**, dust, **benzene**, **naphthalene**, **ammonia**, hydrogen sulfide, and nitrogen oxides, natural gas has such impurities as hydrogen sulfide, water, **condensate**, and dust; while there is hydrogen sulfide, water, **diolefin**, and **raffinate** in LPG.

Quality Indicators of City Gas

To ensure gas appliances to work within the allowed range of applications, and to improve the standardization of gas, the quality indicators of city gas shall meet the following requirements:

(1) The heating value of city gas and **fluctuation** of its components shall comply with the requirements for city gas interchange;

(2) Fluctuation range of city gas different from the reference gas should be adopted according to current "city gas classification and basic characteristics" (GB/T13611-2006), and left with appropriate leeway.

Odorization of City Gas

In case of leaking in the air, city gas should have an appreciable smell that a normal person with an ordinary sense of smell can detect, to allow people to alert about the gas leak and to take timely measures to eliminate hidden dangers. When the odor of city gas itself can not be effectively detected and significantly different from other odors in the everyday environment, **odorant** should be supplemented. The **odorant** currently used in China is mainly **tetrahydrothiophene** (THT), also known as **thiophene**, which is a colorless, non-toxic, non-corrosive transparent oily liquid with a foul odor. It is the most stable compound used for gas odorization, and has a lot of advantages like strong oxidation resistance, stable chemical properties, no pollution to the air, etc., compared with other odorants such as **thiol**s and **thioethers**.

Questions

1. What requirements should city gas meet before being supplied to the users?
2. What are the main impurities in natural gas?
3. What is the **odorant** currently used in China?

Part 4　Translating

Read the following passage about the production of city gas and then translate it into Chinese.

Production of City Gas

Natural gas produced from the wellhead or separated from the field separator contains various amounts of heavier hydrocarbons in the liquid phase under atmospheric conditions, as well as other non-hydrocarbon gases like water vapor, sulfur compounds (e.g. hydrogen sulfide), carbon dioxide, nitrogen and **helium**, generally not suitable for direct use for most users. Most of them need to be treated to remove undesirable components (such as hydrogen sulfide, water vapor) before being used as a commodity for urban users.

LNG is methane-based liquid hydrocarbon compounds liquefied from natural gas. Typically at the normal pressure, natural gas becomes liquid at about $-162\ ℃$, to facilitate transport and storage.

CNG is high-pressure compressed natural gas commodity, whose main component is also methane. It has good anti-explosion performance and good combustion performance with less greenhouse gases and other harmful substances in its combustion products. In addition, its production cost is lower, and thus it is a promising high-quality alternative fuel for vehicles.

LPG is a kind of fuel gas obtained as a by-product in the process of oil or gas **extraction** or oil refining. It is one of the main sources of urban gas in China, and can be divided into two types according to its sources: refinery LPG and oil & gas field LPG. The former is obtained during the secondary processing in **refineries**, mainly composed of propane, propene, butane and butene. The latter is obtained during the natural gas processing, mainly composed of propane or/and butane, but there are no **olefin**s contained.

Part 5　Vocabulary

pillar['pɪlə]　　　　　　　　　　　　*n.* 柱子，柱形物；栋梁；墩
catering['keɪtərɪŋ]　　　　　　　　　*n.* 给养；提供饮食及服务
venting['ventɪŋ]　　　　　　　　　　*n.* 排气；通气

Module 1 An Introduction to City Gas

	v. 排放（vent 的现在分词）
peaking['piːkɪŋ]	*n.* 剧烈增加；脉冲修尖；求峰值
pharmaceutical[ˌfɑːməˈsuːtɪk(ə)l;-ˈsjuː-]	*adj.* 制药（学）的
	n. 药物
pesticide['pestɪsaɪd]	*n.* 杀虫剂
impurities[ɪmˈpjuərɪtɪs]	*n.* 不纯；不洁；杂质
rationality[ˌræʃəˈnælətɪ]	*n.* 合理性；合理的行动
tar[taː]	*n.* 焦油；柏油；水手
benzene['benziːn]	*n.* 苯
naphthalene['næfθəliːn]	*n.* 卫生球；臭樟脑；萘
ammonia[əˈməʊnɪə]	*n.* 氨
condensate['kɒnd(ə)nseɪt]	*n.* 冷凝物；浓缩物
	adj. 浓缩的
diolefin[dɪˈəʊləfɪn]	*n.* 二烯 (=diene)
raffinate['ræfɪneɪt]	*n.* 残油液；剩余液
fluctuation[ˌflʌktʃuˈeɪʃ(ə)n;-tjuː-]	*n.* 起伏，波动
odorization[ˌəʊdə-raɪˈzeɪʃən]	*n.* 加臭
odorant['əʊd(ə)r(ə)nt]	*n.* 添味剂，臭味剂
tetrahydrothiophene[ˌtetrəˌhaɪdrəˈθaɪəfiːn]	*n.* 四氢噻吩
thiophene['θaɪəfiːn]	*n.* 噻吩
thiol['θaɪəʊl]	*n.* 硫醇
thioethers[ˌθaɪəʊˈiːθə]	*n.* 硫醚
extraction[ɪkˈstrækʃn]	*n.* 取出；抽出；拔出；抽出物
helium['hiːlɪəm]	*n.* 氦（符号为 He）
refinery[rɪˈfaɪnərɪ]	*n.* 精炼厂；提炼厂；冶炼厂
olefin['əʊlɪfɪn]	*n.* 烯烃

Module 2

City Gas Network

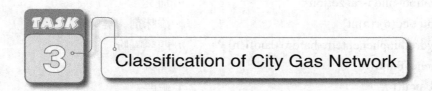

Part 1 Listening

Listen to the following passage and fill in the blanks with proper words.

Classification of Gas Pipeline

First of all, the city gas pipeline 1) _____ the gas **transmission** pressure can be divided into high pressure A and B gas pipeline, 2) _____ pressure A and B gas pipeline, medium pressure A and B gas pipeline, and low pressure gas pipeline. Secondly, according to the use the city gas pipeline is 3) _____ into gas 4) _____ pipeline, gas distribution pipeline, **building service pipeline**, 5) _____ gas pipeline and gas pipeline of 6) _____ enterprises. Thirdly, the city gas pipeline is divided into the underground gas pipeline and 7) _____ gas pipeline according to the 8) _____ mode. Fourthly, the city gas pipeline, according to the 9) _____ of pipe network, is divided into 10) _____ **pipeline network**, **branched pipeline network** and **circular** and branched pipeline network.

Part 2   Speaking

1. Useful sentences

(1) The gas pipelines with various **pressure levels** should be connected through the **pressure regulating device**s in gas transmission and distribution system.

(2) Residential and small commercial consumers generally are supplied by low pressure pipeline directly.

(3) Through the **regulator station** high pressure, gas can be sent into the medium pressure pipeline and the large factories and enterprises whose processes require high pressure gas.

(4) High pressure A gas transmission pipeline is usually the long distance pipeline that is going throughout the provinces, regions or connects the towns, and sometimes it also **constitutes** the outer ring network of the city transmission and distribution pipeline network system.

(5) In city gas pipeline network system, the trunk pipeline with various pressure levels should be connected into the ring network.

(6) Gas transmission pipeline is mainly used for conveying city gas, generally being high pressure, sub-high pressure gas trunk pipeline, or for pipeline connection of large **industrial** enterprises.

(7) Gas distribution pipeline includes the distribution pipelines of the **block** and courtyard, generally being medium pressure and low pressure gas pipeline.

(8) Indoor gas pipeline refers to the pipeline that conveys gas to indoor user through the **total valve** of building service pipeline of the user pipeline and assigns to each of fuel gas appliances.

2. Oral practice

Dialogue

A: Good morning! Professor Wang. Nice to meet you!

B: Morning! Nice to meet you, too!

A: I heard you had recently attended the lecture about the information management technology of gas pipeline network.

B: Very impressive. I have learned a lot.

A: Great! I am writing a paper about it recently. May I ask you some questions?

B: OK, no problem.

A: What kinds of information management technology are there in the city gas industry at present?

B: SCADA system, GIS, MIS and so on.

A: Can you explain them?

B: OK. SCADA system is supervisory control and data **acquisition** system. GIS is geographic information system, and MIS is management information system.

A: Application of information management technology in the gas industry is becoming more and more important.

B: Yes, especially SCADA system. It can be applied in city gas transmission and distribution system, power system, water supply system, petroleum and chemical industry system.

A: Thanks a lot, I learn a lot from you today.

B: It's my pleasure.

Part 3 Reading

City Gas Pipeline Network System

City gas pipeline network system, according to the difference of the adopted pipeline network pressure levels, can be divided into single stage network system, two-stage network system, three-stage network system and **multi-stage network system**. Single stage network system **refers to** the pipeline network system that consists of the pipeline of only one kind of **pressure rating**, generally containing three single stage networks-low pressure, medium pressure A and medium pressure B. Two-stage network system **refers to** the pipeline network system that consists of the pipeline of two kinds of pressure ratings, generally consisting of low pressure and medium pressure two-stage pipeline. Three-stage network system refers to the gas transmission and distribution system that consists of the pipeline of three kinds of pressure ratings, generally consisting of low pressure, medium pressure and high (or sub-high) pressure three-stage pipeline. The pipeline network system that contains more than three kinds of pressure ratings is called multi-stage network system.

In the choice of city gas pipeline network system, we should consider the following main factors:

(1) The gas source conditions, including the types and properties of gas, supply scale and supply pressure, etc.

(2) The present urban situation and development planning, including urban blocks and streets, architectural features, population density, user case (type, quantity, distribution, gas pressure, **gas consumption**, customer percentage, supply principle) and its development planning, etc.

(3) The status of large users and special users, including the number, distribution, gas pressure, gas consumption and the characteristics of the production process, etc.

(4) The measures and capacity for gas storage.

(5) Urban geographical environment and the natural and artificial obstacles, including soil conditions (texture, **corrosion resistance**, temperature, frozen line, etc.), the number and

distribution of bridges, rivers, lakes, railways, etc. the present situation and reconstruction and expansion planning of the underground pipeline and underground architectural structures.

(6) Production and supply situation of required **material of pipe**, **pipe attachment**, pressure regulating equipment, etc.

Questions

1. What is the classification of gas network system according to the difference of pressure levels of the adopted network?
2. Which factors should we consider when choosing gas network system?

Part 4 Translating

（1）在城市燃气管道布线时，必须要考虑城镇发展规划、管道中燃气的压力等基本情况。

（2）室外燃气管道大多采用地下敷设，宜沿城镇道路、人行便道敷设，或敷设在绿化地带内。

（3）室内燃气管道一般为明管敷设，特殊情况有暗埋敷设、暗封敷设。

（4）在城市燃气管道布线时，要依据规范要求确定管道平面与纵断面管位，并进行穿越、跨越障碍物设计。

（5）对于压力大于 1.6MPa 燃气管道通过的地区，应按沿线建筑物密集程度，划分为四个等级，并依据其等级做出相应的管道设计与布置要求。

(6) Gas pipeline whose pressure is no more than 1.6MPa covers the three levels in the pressure rating — sub-high, medium and low pressure, and its main function is to transport gas to the users and to distribute gas to low pressure gas ring network by regulator stations.

(7) Gas pipeline crossing the railway and tramway must adopt the protective sleeve or concrete casing, and vertically cross.

(8) The laying methods of gas pipeline **underwater crossing** river contain the **buried trench laying**, **bare pipe laying**, **pipe jacking laying** and **directional drilling laying**.

(9) The gas pipeline crossing generally contains the **attached-to-bridge crossing**, **pipeline bridge crossing** and **overhead crossing**, etc.

(10) According to the pressure, indoor gas pipeline of residential consumers is divided into two categories — low pressure and medium pressure into the household.

Part 5 Vocabulary

transmission[træns'mɪʃən]　　　　　　*n.* 传输；传送；播送
circular['sɜːkjulə(r)]　　　　　　　　　*adj.* 圆形的；环形的
constitute['kɒnstɪtjuːt]　　　　　　　　*vt.* 组成，构成；建立；任命
industrial[ɪn'dʌstrɪəl]　　　　　　　　　*adj.* 工业的，产业的；供工业用的

block[blɒk]	n. 块；障碍物，阻碍；街区
acquisition[ˌækwɪˈzɪʃ(ə)n]	n. 获得物，获得
building service pipeline	用户引入管
branched pipeline network	枝状管网
circular pipeline network	环状管网
pressure level	压力级制
pressure regulating device	调压装置
regulator station	调压站
total valve	总阀门
multi-stage network system	多级管网系统
refers to	指的是
pressure rating	压力等级
gas consumption	用气量
corrosion resistance	耐腐蚀性能
material of pipe	管材
pipe attachment	管道附件
underground laying	地下敷设
indoor open installation	明管敷设
piping embedment	暗埋敷设
piping concealment	暗封敷设
underwater crossing	水下穿越
buried trench laying	沟埋敷设
bare pipe laying	裸管敷设
pipe jacking laying	顶管敷设
directional drilling laying	定向钻敷设
attached-to-bridge crossing	附桥跨越
pipeline bridge crossing	管桥跨越
overhead crossing	架空跨越

Module 2 City Gas Network

TASK 4 Supply of City Gas

Part 1 Listening

Listen to the following passage and fill in the blanks with proper words.

Storage and Transportation of City Gas

Generally speaking, the 1) _____ used storage methods of natural gas include **gas storage tank**, gas 2) _____ **reservoir, gas pipeline and tube bundle**, 3) _____ natural gas storage. **Gas storage tank**s are divided into low pressure gas storage tanks and high pressure gas storage tanks. The **underground storage** of natural gas usually includes the 4) _____ oil and gas fields storage, **aquifer** 5) _____ **formation storage, salt layer storage** and **cave storage**, etc. 6) _____ **liquefied natural gas storage** generally adopts the method of **low temperature and** 7) _____ **pressure storage**. Other storage methods of natural gas include the storage in low temperature liquefied 8) _____ gas and **solid-state storage,** etc. Transportation 9) _____ of city gas include pipeline transportation, **rail tanker transportation**, **road tanker transportation** and 10) _____ **transportation**.

城市燃气专业英语 04

Part 2 Speaking

1. Useful sentences

(1) The storage modes of natural gas, according to its principles and characteristics, can be roughly divided into low pressure gas storage and high pressure gas storage, etc.

(2) The storage modes of natural gas, according to its storage form, can be divided into the gas-state storage and the liquid-state storage, etc.

(3) The **depleted** oil and gas fields are the best and most reliable gas underground reservoirs.

(4) Solid-state storage of natural gas is to change natural gas at a certain pressure and temperature into the **solid crystalline hydrate** which is stored in the steel storage tanks.

- 17 -

(5) According to the geological structure, gas underground reservoir is divided into depleted oil and gas fields, **aquifer**, salt cave, rock cave and abandoned mine.

(6) There are two types of low pressure gas storage tanks: **low pressure piston-type tank** and **low pressure water-sealed tank**.

(7) High pressure gas storage tank is a kind of volume tank, commonly used in two forms: **horizontal cylindrical tank** and **spherical tank**.

(8) High pressure gas tube bundle storage and long-distance pipeline end storage are the effective ways of balancing hourly **uneven** gas consumption.

(9) Compared with the gas storage facilities on the ground, gas underground reservoir has a series of advantages of large capacity, strong adaptability, good economy, high safety, less occupation area and small environmental impact, etc.

(10) Workshop pipeline system should be indoor open installation whose erecting height should be not less than two meters.

2. Oral practice

Dialogue

A: Good morning! Professor Li. I have been studying natural gas storage by myself, but I meet with some difficulties. May I ask you some questions?

B: Of course!

A: Thank you! I saw a term that I am not familiar with on the book. That is ANG.

B: Hum, ANG is the **adsorption storage of natural gas**, which is one of the storage methods of natural gas.

A: I see. Can you introduce some knowledge for adsorption storage of natural gas (ANG)?

B: Ok. Adsorption storage of natural gas (ANG) is to load the solid adsorbent (such as zeolite, molecular sieve, silica gel, carbon black, activated carbon, etc.) in the storage tank, so as to adsorb natural gas under certain storage pressure (from 3.5MPa to 6.0MPa), and achieve the storage capacity close to compressed natural gas (CNG).

A: What advantages of adsorption storage of natural gas (ANG) are there?

B: The advantages of adsorption storage of natural gas (ANG) are relatively low storage pressure, low requirements of pressure performance for gas storage and filling equipment, low investment costs, and safe and reliable performance, convenient routine operation and maintenance, low operating costs.

A: Now I have some ideas about ANG. Thank you very much!

B: You are welcome.

Part 3 Reading

Gas Supply System of Industrial Enterprises

Gas transmission and distribution system of industrial enterprises is usually constituted by factory service pipelines, plant pipelines, workshop pipelines, factory total regulator stations or workshop pressure regulating devices, **gas metering devices**, safety control devices and **stokehole pipelines**. Industrial enterprise users are generally supplied by urban medium pressure or sub-high pressure pipeline network, the users of small gas consumption and low gas pressure can be directly supplied by low pressure pipeline network, and the choice of the best solution should be decided by the technical and economic analysis results. Large industrial enterprises can lay the special pipeline connected with city gas gate station or long-distance pipeline.

Gas network system of industrial enterprises can be broadly divided into two categories — single stage network system and two-stage network system. In the choice of gas network system of industrial enterprises, we should consider the following main factors:

(1) Gas pressure of city gas distribution network with service pipeline connection.

(2) The **rated pressure** required by gas burners of each gas workshop.

(3) The location of gas workshop distributed in the factory.

(4) Gas consumption and gas scale of workshop.

(5) The relations with other pipelines, management and maintenance conditions and economic effect. As to the gas pipeline wiring in plants, we should follow the following principles:

(1) Gas pipelines used in plants are generally steel pipes.

(2) Gas service pipeline should be laid in gas-using room or gas meter room.

(3) Plant gas pipeline can be laid either underground or overhead.

(4) **Vent pipe** should be set up for the end of plant gas pipeline.

(5) Plant **overhead gas pipeline** system should be as simple and clear as possible, in order to facilitate the construction and installation, operation management and the routine maintenance.

There are two kinds of workshop gas network system: the branched and the circular, the former of which is commonly used, and the latter is only used for particularly important workshop.

Questions

1. What is gas transmission and distribution system of industrial enterprises usually constituted by?

2. Which factors should we consider in the choice of gas network system of industrial enterprises?

Part 4 Translating

(1) 室内燃气管道系统一般由用户引入管、水平干管、立管、用户支管、燃气计量表、用具连接管和燃具所组成。

(2) 按照燃气表的设置方式，室内燃气管道又可分为分散设表和集中设表两类。

(3) 用户引入管是指从室外配气支管与用户室内燃气进口总阀门之间的管道。

(4) 燃气立管不得敷设在卧室或卫生间内。

(5) 住宅内燃气支管不能明设时，可采用暗埋或暗封设置。

(6) Gas branched pipeline through the wall should be installed in the casing.

(7) High-rise building with its large weight and significant settlement is easy to be damaged in the service pipeline.

(8) High-rise building gas **riser** with its long pipeline and significant settlement requires **piers** setting up at its bottom.

(9) **Appliance connecting pipeline** (also called **sagging pipe**) is a vertical pipe section which is connected with **gas appliance**s in a gas branched pipeline.

(10) Material of indoor gas pipeline should adopt the conveying steel pipe for low pressure fluid, and should as far as possible adopt **galvanized steel pipe**.

Part 5 Vocabulary

depleted[dɪˈplɪtɪd]	adj. 耗尽的；废弃的；贫化的
aquifer[ˈækwɪfə]	n. 含水层，地下蓄水层
uneven[ʌnˈiːv(ə)n]	adj. 不均匀的；不平均的
scatter[ˈskætə(r)]	vt. (使) 散开，(使) 分散
	vi. 散开；分散
riser[raɪzə(r)]	n. 立管；起义者；叛乱者
pier[pɪə(r)]	n. 码头，防波堤；桥墩
gas storage tank	储气罐 (储气)
gas underground reservoir	地下储气库 (储气)
gas pipeline and tube bundle	管道和管束 (储气)
salt layer storage	盐矿层 (储气)
cave storage	岩穴 (储气)
liquefied natural gas storage	液化天然气储存
low temperature and atmospheric pressure storage	低温常压储存
depleted oil and gas fields storage	枯竭的油气田 (储气)
aquifer porous formation storage	含水多孔地层 (储气)
solid-state storage	固态储存

Module 2 City Gas Network

rail tanker transportation	铁路槽车运输
road tanker transportation	公路槽车运输
waterway transportation	水路槽船运输
solid crystalline hydrate	固体结晶水化物
low pressure piston-type tank	低压干式罐
low pressure water-sealed tank	低压湿式罐
horizontal cylindrical tank	卧式圆筒形罐
spherical tank	球形罐
adsorption storage of natural gas(ANG)	天然气的吸附储存
gas metering device	用气计量装置
stokehole pipeline	炉前管道
rated pressure/normal operating pressure	燃气额定压力
vent pipe	放散管
overhead gas pipeline	架空燃气管道
horizontal main pipeline	水平干管
user branched pipeline	用户支管
gas meter	燃气计量表
appliance connecting pipeline	用具连接管
sagging pipe	下垂管
gas appliance	燃具
galvanized steel pipe	镀锌钢管

- 21 -

TASK 5 Gas Network Peak Shaving

Part 1   Listening

Listen to the following passage and fill in the blanks with proper words.

Supply and Demand Balance of City Gas

City gas consumption 1) _____ with time, the monthly, daily and hourly are not the same, but the **gas sendout** of gas source generally changes little, 2) _____ long-distance gas transmission pipeline. 3) _____, unbalance between gas sendout and gas consumption often occurs. In order to ensure supply according to the 4) _____ of customers, we must solve the 5) _____ problem of gas sendout and gas consumption. To solve the problem of the monthly (quarterly), daily or hourly 6) _____ gas consumption we can adopt different storage methods, 7) _____ **underground storage**, gas storage tank, 8) _____ storage and gas transmission pipeline end storage, etc. **Underground storage** is mainly used to 9) _____ the **seasonal uneven gas consumption**, and gas storage tank, **liquid-state storage** and gas transmission pipeline end storage are mainly used to 10) _____ the unbalance problem of hourly gas consumption.

Part 2   Speaking

1. Useful sentences

(1) In order to solve the **contradiction** between even supply and uneven consumption, we must take the appropriate methods to make the supply and demand balance of gas transmission and distribution system.

(2) The primary role of the various gas storage facilities is to store the excess gas when gas consumption is less than gas sendout, and to make up for the lack when gas consumption is more than gas sendout.

(3) During the gas-using peak, the supply fails to meet the demand, while during the low peak, the supply exceeds the demand.

Module 2　City Gas Network

(4) City gas belongs to the downstream of the whole natural gas system, long distance transmission pipeline belongs to the midstream, and the **exploitation** and **purification** of natural gas belongs to the upstream.

(5) The low temperature liquid-state storage is suitable for long distance transportation, and usually storage capacity is very large in consideration of the economical efficiency.

(6) Gas storage tank refers to the equipment specially used to store gas.

(7) The long-distance gas transmission pipeline or high pressure gas pipeline of urban outer ring storage is one of the most economical methods used for storing gas, and also one of the most commonly used methods at home and abroad.

(8) The **buffered user** of city gas supply refers to the interruptible gas users who can make the gas fluctuations in a year minimum and the non-peak period gas users.

2. Oral practice

Dialogue

A: Good afternoon! Professor Su.

B: Afternoon! Nice to meet you!

A: I'm afraid I have to take you some time. I've got some questions to ask you. Can you tell me why the unbalance of supply and demand of city gas exists?

B: No problem. City gas consumption varies with time, but the gas sendout of gas source generally changes little. Therefore, unbalance between gas sendout and gas consumption often occurs.

A: In order to ensure supply according to the demand of users, we must solve the unbalance problem.

B: Yes, at present, people try their best to adopt various methods to solve the unbalance problem.

A: Can you briefly explain to me the methods of adjusting the unbalance of supply and demand of city gas?

B: Hum, the first method is to change the manufacturability of gas source and set up the **flexible** gas source. To this method we must consider the difficulty level of operation and stop of gas source, the possibility and variation **amplitude** of production load variation of gas source. At the same time, we should consider the safety reliability and the technical and economic **rationality** of gas supply.

A: I see. In addition to this method, are there other methods?

B: Yes. There are another two methods. One is to use the buffered users to adjust the unbalance problem. This method is mainly to use some large industrial enterprises and the boiler room as the buffered users, and we can use this method to balance the seasonal uneven gas consumption and part of the daily uneven gas consumption. The other is to use gas storage facilities to adjust the unbalance problem.

A: Thanks a lot, I learn a lot from you today.

B: It's my pleasure.

Part 3　Reading

Volume Calculation of Gas Storage

Determination of the volume of gas storage tank

The calculation of the required gas storage tank capacity of city gas transmission and distribution system, according to the gas source and whether the gas transmission supply can meet the requirement of the daily gas consumption, is divided into two kinds of working conditions:

(1) When gas supply can change according to the daily gas consumption required, gas storage tank capacity should be calculated according to the balance condition of gas supply and demand of 24 hours on the design day in **design month**.

(2) When gas supply can't change according to the daily gas consumption required, gas storage tank capacity should be calculated according to the balance condition of gas supply and demand of the average 168 hours of a week in design month.

General specific steps to determine the volume of gas storage tank are as follows:

(1) To determine the **maximum uneven factor of monthly consumption**.

(2) To determine the **uneven factor of daily consumption** and the **maximum uneven factor of daily consumption**.

(3) To determine the **uneven factor of hourly consumption**.

(4) To calculate the **annual average daily gas consumption** of the residential and commercial users.

(5) To calculate the **monthly average daily gas consumption** of the residential and commercial users.

(6) To calculate the hourly gas consumption of any single day within a week of the residential and commercial users.

(7) To calculate the hourly gas consumption of the industrial users.

(8) To calculate the accumulated value of gas consumption.

(9) To calculate the accumulated value of gas sendout.

(10) To calculate the hourly gas storage capacity.

(11) To **respectively** calculate the related values of the above by using the list method under two conditions of 24 hours or per hour in a week.

(12) The required gas storage volume is equal to the sum of the absolute value of the maximum and minimum hourly gas storage capacity.

The calculation of gas storage capacity of long distance gas transmission pipeline end

Gas storage of long distance gas transmission pipeline end is to use the pipeline between the last compressor station of long-distance gas transmission system and city gate station to store gas. Its storage capacity is the difference between the gas storage volume in this section of gas transmission pipeline at the end and beginning of gas storage, and it can be calculated approximately according to the following steps:

(1) According to the required minimum supply pressure of the end user and the normal gas transmission capacity to determine the starting pressure $p_{1\min}$ at the beginning of gas storage:

$$p_{1\min} = \sqrt{p_{2\min}^2 + \frac{9.053 Q_V^2 \lambda Z_{\min} STL}{d^5 \times 10^7}}$$

Symbol description

$p_{1\min}$ — starting absolute pressure of gas pipeline at the beginning of gas storage, MPa;
$p_{2\min}$ — terminal absolute pressure of gas pipeline at the beginning of gas storage, MPa;
Q_V — flow of gas ($p_0 = 0.101325$ MPa, $T_0 = 293$ K), m³/d;
λ — hydraulic friction factor;
$Z_{\min}$ — average compressibility factor in gas pipeline at the beginning of gas storage;
S — relative density of gas;
T — average temperature of gas in gas pipeline, K;
L — length of gas pipeline, km;
d — inner diameter of gas pipeline, cm.

(2) According to the maximum working pressure of the compressor station or the allowable pressure of the pipeline strength and the normal gas transmission capacity to determine the end pressure $p_{2\max}$ at the end of gas storage by the following formula:

$$p_{2\max} = \sqrt{p_{1\max}^2 - \frac{9.053 Q_V^2 \lambda Z_{\max} STL}{d^5 \times 10^7}}$$

Symbol description

$p_{2\max}$ — terminal absolute pressure of gas pipeline at the end of gas storage, MPa;
$p_{1\max}$ — starting absolute pressure of gas pipeline at the end of gas storage, MPa;
$Z_{\max}$ — average compressibility factor in gas pipeline at the end of gas storage.

(3) To determine the **average absolute pressure** $p_{m.\min}$ at the beginning of gas storage:

$$p_{m.\min} = \frac{2}{3}\left(p_{1\min} + \frac{p_{2\min}^2}{p_{1\min} + p_{2\min}}\right)$$

Symbol description

$p_{m.\min}$ — average absolute pressure in gas pipeline at the beginning of gas storage, MPa.

(4) To determine the average absolute pressure $p_{m.\max}$ at the end of gas storage:

$$p_{m.\max} = \frac{2}{3}\left(p_{1\max} + \frac{p_{2\max}^2}{p_{1\max} + p_{2\max}}\right)$$

Symbol description

$p_{m.max}$ — average absolute pressure in gas pipeline at the end of gas storage, MPa.

(5) To determine the gas storage capacity V of gas transmission pipeline:

$$V = \frac{V_c T_0}{p_0 T} \left(\frac{p_{m.max}}{Z_{max}} - \frac{p_{m.min}}{Z_{min}} \right)$$

Symbol description

V — gas storage capacity of gas pipeline ($p_0 = 0.101325$ MPa, $T_0 = 293$ K), m³;

V_c — geometric volume of gas pipeline, m³.

Questions

1. What are the general specific steps to determine the volume of gas storage tank ?

2. What are the steps of approximate calculation of gas storage capacity of long distance gas transmission pipeline end ?

Part 4 Translating

(1) 为了保证按用户要求不间断地供应燃气，必须考虑燃气的供应与使用的平衡问题。

(2) 在调节燃气供需平衡时，通常是由上游供气方解决季节性供需平衡，下游用气城镇解决日供需平衡。

(3) 当用气城市距天然气产地不太远时，可采用调节气井供应量的办法平衡部分日不均匀性用气。

(4) 根据工业企业、居民生活及商业的用气量和用气工况，制订调度计划来调整供气量。

(5) 地下储气库储气量大，造价和运行费用低，可用来平衡季节不均匀用气和一部分日不均匀用气。

(6) After the arrival of gas supply of long-distance gas transmission pipeline, both sides of supply and demand should be clear with each other in the responsibilities of **peak shaving** and safe gas supply.

(7) Throughout city gas transmission and distribution system, it is necessary to solve the problem of peak shaving according to the overall situation, to achieve the optimization of the natural gas system and economic **rationality**.

(8) Liquefied natural gas which can be vaporized conveniently and adjusted extensively in terms of its load range, is suitable for adjusting the various uneven gas consumption.

(9) The metal storage tank is essential to guarantee the normal operation of gas pipeline network.

(10) With the growing popularity of natural gas, high-pressure spherical storage tanks are commonly used as the peak-shaving gas source in large and medium city gate stations.

Part 5 Vocabulary

contradiction[ˌkɒntrəˈdɪkʃ(ə)n]	n. 矛盾；否认，反驳
exploitation[ˌeksplɔɪˈteɪʃ(ə)n]	n. 开发；利用；剥削；广告推销
purification[ˌpjuərɪfɪˈkeɪʃn]	n. 净化；洗净；提纯
fluctuation[ˌflʌktʃuˈeɪʃn]	n. 波动，涨落，起伏
flexible[ˈfleksəbl]	adj. 灵活的；易弯曲的；柔韧的
amplitude[ˈæmplɪtjuːd]	n. 振幅；广大，广阔，充足
rationality[ˌræʃəˈnælətɪ]	n. 合理性；（复数）合理的行动
respectively[rɪˈspektɪvlɪ]	adv. 各自地；各个地；分别地
peak shaving	调峰
gas sendout	供气量
underground storage	地下储气
seasonal uneven gas consumption	季节用气不均衡性
liquid-state storage	液态储气
buffered user	缓冲用户
design month	计算月
maximum uneven factor of monthly consumption	月高峰系数
uneven factor of daily consumption	日不均匀系数
maximum uneven factor of daily consumption	日高峰系数
uneven factor of hourly consumption	小时不均匀系数
annual average daily gas consumption	年平均日用气量
monthly average daily gas consumption	月平均日用气量
average absolute pressure	平均绝对压力

Module 3

Process in Gas Transmission and Distribution

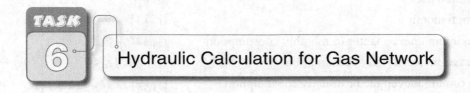

Part 1 Listening

Listen to the following passage and fill in the blanks with proper words.

The Computer Application in Hydraulic Calculation

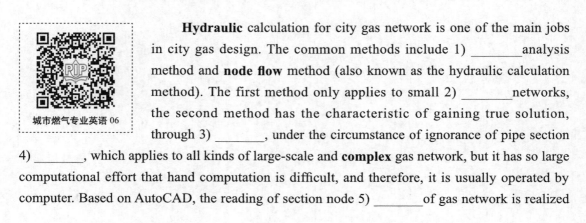

Hydraulic calculation for city gas network is one of the main jobs in city gas design. The common methods include 1) _____ analysis method and **node flow** method (also known as the hydraulic calculation method). The first method only applies to small 2) _____ networks, the second method has the characteristic of gaining true solution, through 3) _____, under the circumstance of ignorance of pipe section 4) _____, which applies to all kinds of large-scale and **complex** gas network, but it has so large computational effort that hand computation is difficult, and therefore, it is usually operated by computer. Based on AutoCAD, the reading of section node 5) _____ of gas network is realized

Module 3 Process in Gas Transmission and Distribution

by VB secondary development technology and 6) _____ calculation drawing of gas network is **automatically** 7) _____, and the hydraulic calculation software for city gas network can be programmed by VC++ visual programming language.

Part 2 Speaking

1. Useful sentences

(1) With the rapid development and popularization of computer technology, the difficulty of large workload with hand computation is overcome.

(2) Hydraulic calculation is one of the main jobs in city gas design.

(3) Determine the pipeline investment and metal **consumption** by the hydraulic calculation, and guarantee the work reliability of pipe network.

(4) In the practical engineering, flow producing **local resistance** is often in the rough region of **turbulent** flow.

(5) Total resistance of pipe section contains **frictional resistance** and local resistance.

(6) Basically, pipe network can be divided into branched network and **looped network**.

(7) **Indoor gas pipeline** is the pipeline from **service pipe** to gas appliances in the end of pipeline.

(8) **Distribution flow** is the flow which is distributed to the users along pipe section directly.

(9) Any pipe network graphic is the geometric figure connected by some nodes and pipe sections.

(10) A powerful tool for expressing pipe network graphic properties is **matrix**, including incidence matrix and relation of nodes matrix.

2. Oral practice

Dialogue

Setting: Two students (A and B) are talking about hydraulic calculation for city gas network. Read the dialogue and answer the following questions.

A: Hydraulic calculation for city gas network is one of the main jobs in city gas design.

B: You're right! At design time, that gas network not only meets the needs of use, but also saves investment is required.

A: In addition, for the running gas network, the reasonable production scheduling and accident simulation for gas network can be guaranteed, and the disposal plans and emergency treatment of gas network accidents can be established.

B: We have just talked about the principles of designing. So, what is its specific tasks?

A: Simply speaking, the task of hydraulic calculation for city gas network is to determine the pipe **diameters**, according to the **design flow** rate and **allowable pressure drop** of gas.

B: What are the conditions of formula derivation?

A: Under normal conditions and in a short period of time, gas flow in gas pipeline can be regarded as **steady** flow.

B: What else?

A: There is a **hypothesis**, which is that gas flow is **isothermal** flow, so λ and T can be a **constant**; considering the pressure change is not too big, so **variable** Z also can be a constant.

B: Oh. Let me see. In the actual computation, using the formula directly may be tedious, right?

A: Yes! Usually, for simplified calculation, the derived hydraulic computational formula will be drawn into hydraulic calculation drawings.

B: Is it necessary to do the **parameter modification**?

A: Right! Because calculation drawings are drawn under the specific parameter of gas, it is necessary to do the parameter modification when there are differences between actual parameter and parameter of calculation drawings.

B: All right. Thanks very much for your explanation. I have known a lot about hydraulic calculation for city gas network.

A: It's my pleasure.

Part 3 Reading

Hydraulic Calculation for City Gas Pipeline

It is a problem to do the hydraulic computation accurately, which is related to economical efficiency and reliability of transmission and distribution system, it is also an important part of planning and design of city gas.

Hydraulic Computational Formulas and Diagrams of City Gas Pipeline

1) Hydraulic computational formulas of low-pressure gas pipeline

When gas is considered under conditions of different material pipelines and different flow states, the friction loss of unit length of low-pressure gas pipeline is calculated specifically using the following formulas:

$$\frac{\Delta p}{L} = 1.13 \times 10^{10} \frac{Q}{d^4} v \rho \frac{T}{T_0} \qquad \text{(\textbf{Laminar} region, } Re<2100)$$

$$\frac{\Delta p}{L} = 1.88 \times 10^6 \left(1 + \frac{11.8Q - 7 \times 10^4 dv}{23Q - 1 \times 10^5 dv}\right) \frac{Q^2}{d^5} \rho \frac{T}{T_0} \qquad \text{(\textbf{Critical} region, } Re=2100\sim3500)$$

$$\frac{\Delta p}{L} = 6.89 \times 10^6 \left(\frac{K}{d} + 192.2 \frac{dv}{Q}\right)^{0.25} \frac{Q^2}{d^5} \rho \frac{T}{T_0} \qquad \text{(\textbf{Turbulent} region, } Re>3500, \text{Steel tube, Plastic pipe})$$

Module 3 Process in Gas Transmission and Distribution

$$\frac{\Delta p}{L} = 6.39 \times 10^6 \left(\frac{1}{d} + 5158\frac{dv}{Q}\right)^{0.284} \frac{Q^2}{d^5} \rho \frac{T}{T_0} \quad \text{(\textbf{Turbulent} region, } Re>3500, \text{ Cast iron pipe)}$$

Symbol description

  Re—reynolds number;
  Δp—friction loss of gas pipeline, Pa;
  L—calculated length of gas pipeline, m;
  Q—design flow of gas pipeline, Nm^3/h;
  d—inner diameter of gas pipeline, mm;
  ρ—gas density, kg/m^3;
  T—gas temperature used in the design, K;
  T_0—value as 273.15, K;
  v—kinematic viscosity ($p_0 = 101.325$ kPa, $T_0 = 0\,°C$), m^2/s;
  K—equivalent absolute roughness of inner surface of gas pipeline, mm.

2) Hydraulic computational formulas of high and medium-pressure gas pipeline

When the friction coefficient of gas pipeline is calculated by hand, according to the different material of gas pipeline, the friction loss of unit length of high and medium-pressure gas pipelines should be calculated as follows:

$$\frac{p_1^2 - p_2^2}{L} = 1.4 \times 10^9 \left(\frac{K}{d} + 192.2\frac{dv}{Q}\right)^{0.25} \frac{Q^2}{d^5} \rho \frac{T}{T_0} \quad \text{(Steel tube, Plastic pipe)}$$

$$\frac{p_1^2 - p_2^2}{L} = 1.3 \times 10^9 \left(\frac{1}{d} + 5158\frac{dv}{Q}\right)^{0.284} \frac{Q^2}{d^5} \rho \frac{T}{T_0} \quad \text{(Cast iron pipe)}$$

Symbol description

  p_1— starting pressure (**absolute pressure**) of gas pipeline, kPa;
  p_2— terminal pressure (absolute pressure) of gas pipeline, kPa;
  L— calculated length of gas pipeline, km.

3) Hydraulic computational diagrams of city gas pipeline

If pressure or tubular product is different, hydraulic computational formulas are also different, which will have different hydraulic computational charts. In addition, because the different types of fuel gas are very different in density and viscosity, etc., the calculation chart is also different.

Three influence factors deciding hydraulic computational charts are mainly different types of gas, different pressure level of pipeline and different tubular product. Different combinations of the three get different hydraulic calculation chart.

4) Additional **pressure head**

Because of different density of air and gas, when the elevation difference exists at the beginning and the end of the pipeline, an additional pressure head is generated in the gas pipeline.

There is a large variation in the height difference between the beginning and the end for some pipe section, including the low pressure distribution pipe and the low pressure gas pipeline inside the building, and additional pressure head must be counted in.

Computational formula as follows:

$$\Delta p = g(\rho_a - \rho_g)\Delta H$$

Symbol description

Δp — Additional pressure head, Pa;

ρ_a — Air density, $\rho_a = 1.293$, kg/m^3;

ρ_g — Gas density, kg/m^3;

g — Acceleration of gravity, $g=9.81$ m/s^2;

ΔH — The elevation difference between the end and the beginning of the pipeline.

When gas in pipeline goes up, it will produce a kind of lift force, on the contrary, it will increase resistance ($\rho_a > \rho_g$).

When gas in pipeline goes down, it will produce a kind of lift force, on the contrary, it will increase resistance ($\rho_a < \rho_g$).

5) Calculation of local pressure loss

When gas flows in some pipeline fittings, including **t-branch** pipe, **elbow** pipe, different diameters and special pipes, valve, it will produce additional pressure loss inevitably, because of rapid change of geometric boundary, with the change of **airflow** direction and fracture surface, which will cause the movement of gas to be disturbed.

In general, for city gas network, local pressure loss is 5% ~ 10% of **linear loss**.

For the indoor gas pipelines and those in the area of factory or station, there is a large proportion of local pressure loss, because of many pipeline fittings. There are two methods to calculate: formula method and **equivalent length** method.

Formula method as follows:

$$\Delta p = \Sigma \xi \frac{\upsilon^2}{2}\rho$$

Symbol description

Δp — Loss of local resistance of pipe segment, Pa;

$\Sigma \xi$ — The sum of the local resistance coefficients in the calculated pipe segment;

υ — Gas flow rate, m/s.

Equivalent length method as follows:

$$\Delta p = \Sigma \xi \frac{\upsilon^2}{2}\rho = \lambda \frac{L_d}{d}\frac{\upsilon^2}{2}\rho$$

$$L_d = \Sigma \xi \frac{d}{\lambda}$$

Module 3 Process in Gas Transmission and Distribution

Symbol description

λ — Hydraulic **friction factor**;

L_d — The equivalent length of local friction resistance of pipe segment, m;

d — The inner diameter of gas pipeline, m.

Branched Network

Hydraulic calculation procedures:

(1) To number the nodes of gas network and pipelines.

(2) To determine the design flow of each section of pipeline network according to pipeline drawing and gas consumption.

(3) To choose the trunk pipeline of branched network, and to determine the allowed pressure drop on the unit length of pipeline according to the given allowed pressure drop.

(4) To select pipe diameter preliminarily according to the design flow of the pipe section and the allowed pressure drop of unit length.

(5) According to the selected standard pipe size, to reversely calculate the actual frictional pressure drop and local pressure drop, and to calculate the total pressure drop.

(6) To check the computed result. If total pressure drop is not in excess of the allowable value and approaches it, then the computed result is considered qualified, otherwise, the pipe diameter should be properly changed, until the total pressure drop is less than and approaches the allowable value as soon as possible.

Loop Pipe Network

1) Flow **regularity**

(1) Equation of **continuity** of node.

(2) **Closure** difference of looped pressure drop is equal to zero.

2) Hydraulic calculation procedures

Usually, the method of equation set to solve pipe section, loop and node is adopted in hydraulic computation of loop pipe network. It is **simultaneous** to solve the equation of pressure drop, equation of continuity and energy equation.

(1) To draw planar graph of pipe network, number the nodes, pipe sections and loop network, and mark length of pipeline, concentrated load, location of gas source or regulator station.

(2) To compute distribution flow of every pipe section of pipe network.

(3) According to the principle of gas flowing from site of gas supply to zero point along the shortest path, to study out flow direction of gas in every pipe section. The flow direction of gas is always far away from the site of gas supply, not reversing flow.

(4) To begin with zero point and calculate **transit flow** of pipe section one by one.

(5) To calculate the design flow of every section of pipe network.

(6) To select pipe diameter preliminarily:

① To determine the allowed pressure drop of unit length, according to the allowed pressure drop of pipe network and the calculated length of from site of gas supply to zero point.

② To select pipe diameter preliminarily, according to the allowed pressure drop of unit length.

③ To put pipe diameter into the specification series.

(7) To **preliminary** calculate the pressure drop of every pipe section of pipe network and the closure difference of pressure drop per loop.

(8) To make compensating computation of pipe network, which is to solve the corrected flow for per loop, so that the algebraic sum of pressure drop of all closed-loop network equals or approaches to zero, reaching the allowed error range of project.

(9) To verify the overall pressure drop, if requirements are met, then the calculation is finished; otherwise, to repeat (6), (7), (8).

Part 4　Translating

(1) 在进行室内燃气管道水力计算时，首先应根据建筑物的平面图和剖面图来选定和布置用户燃气用具，画出管道系统图。

(2) 利用同时工作系数法计算确定各管段的计算流量。

(3) 自引入管到各燃具之间的压降，其最大值为系统的压力降。

(4) 管段按顺序编号，凡管径变化或流量变化处均应编号，并标上各计算管段的实际长度。

(5) 根据计算流量设定各管段的管径。

(6) Hydraulic computation drawing is chosen according to kind, density and **kinematic** viscosity of gas, and the value of pressure drop of unit length of pipe section is confirmed.

(7) The additional pressure of every pipe section is calculated.

(8) The actual pressure loss of every pipe section is calculated.

(9) When total pressure drop of indoor gas pipeline is calculated, the **design pressure** drop of **manufactured** gas is within 80 ~ 100Pa generally.

(10) When a comparison is made between total pressure drop of indoor gas pipeline and allowable pressure drop, if inappropriate, it is necessary to adjust the pipe diameter of individual pipe section.

Part 5　Vocabulary

hydraulic[haɪˈdrɔːlɪk]	adj. 水力的；水力学的
branched[brɑːntʃt]	adj. 分支的；分岔的；枝状的
iteration[ˌɪtəˈreɪʃn]	n. 迭代
looped[luːpt]	adj. 环状的；成圈的
node[nəʊd]	n.（计算机网络的）节点
flow[fləʊ]	n. 流量；流动

Module 3 Process in Gas Transmission and Distribution

complex['kɒmpleks]	*adj.* 复杂的；复合的
automatically[ˌɔːtəˈmætɪklɪ]	*adv.* 自动地；机械地；无意识地
consumption[kənˈsʌmpʃn]	*n.* 消耗；消耗量
laminar[ˈlæmɪnə]	*adj.* 层流的
	n. 层流
resistance[rɪˈzɪstəns]	*n.* 阻力
turbulent[ˈtɜːbjəl(ə)nt]	*adj.* 紊流的；湍流的
	n. 紊流；湍流
matrix[ˈmeɪtrɪks]	*n.* 矩阵
diameter[daɪˈæmɪtə(r)]	*n.* 直径
steady[ˈstedɪ]	*adj.* 稳定的；稳态的
hypothesis[haɪˈpɒθəsɪs]	*n.* 假设；假说；假定
isothermal[ˈaɪsəʊˈθɜːməl]	*adj.* 等温的；恒温的
constant[ˈkɒnstənt]	*n.* 常数；常量
variable[ˈveərɪəbl]	*adj.* 变化的；可变的
parameter[pəˈræmɪtə(r)]	*n.* 参数
modification[ˌmɒdɪfɪˈkeɪʃn]	*n.* 修正；修改
critical[ˈkrɪtɪkl]	*adj.* 临界的
elbow[ˈelbəʊ]	*n.* 弯头；弯管
airflow[ˈeəfləʊ]	*n.* 气流
equivalent[ɪˈkwɪvələnt]	*n.* 当量；等效
regularity[ˌregjuˈlærətɪ]	*n.* 规律；规律性
continuity[ˌkɒntɪˈnjuːətɪ]	*n.* 连续；连续性
closure[ˈkləʊʒə(r)]	*n.* 闭合
simultaneous[ˌsɪmlˈteɪnɪəs]	*adj.* 联立的
preliminary[prɪˈlɪmɪnərɪ]	*adj.* 初步的
kinematic[ˌkɪnɪˈmætɪk]	*adj.* 运动的；运动学的
manufactured[mænjuˈfæktʃəd]	*adj.* 人造的；人工的
local resistance	局部阻力
frictional resistance	摩擦阻力
looped network	环状管网
indoor gas pipeline	室内燃气管道
service pipe	引入管
distribution flow	途泄流量
design flow	计算流量
allowable pressure drop	允许压降
absolute pressure	绝对压力
pressure head	压头

t-branch	三通
linear loss	沿程损失
equivalent length	当量长度
friction factor	摩擦阻力系数
transit flow	转输流量
design pressure	计算压力

Module 3 Process in Gas Transmission and Distribution

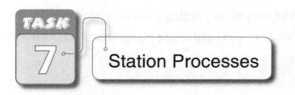

Task 7 Station Processes

Part 1 Listening

Listen to the following passage and fill in the blanks with proper words.

Natural Gas Storage and Distribution Station Process

The processes of two-stage pressure regulation, gas storage of high-pressure and gas supply of sub-high pressure are adopted in the natural gas storage and **distribution** station. Pressure of high-pressure main pipe of natural gas, which flows from gas 1) _____ station to factory storage and distribution station, is 0.5 ~ 1.5MPa, this pressure turns to be 0.3MPa after it is regulated in the storage and distribution station, which flows in the network of gas supply of sub-high pressure for industrial and 2) _____ use. When the storage and distribution station is low-peak load, natural gas getting in the station by measuring and regulating is supplied to pipe network of sub-high pressure 3) _____, and at the same time, it is made to take the 4) _____ to gas storage tank. When gas pressure getting in the station is lower than that of gas storage tank, it is needed to use the 5) _____ to pressure up to that of gas storage 6) _____, and then the gas is made to take the **aeration** to gas storage tank. In general, pressure of gas storage tank is 0.8 ~ 1.5MPa. When users are in high-peak load, natural gas must be let out from gas storage tank and get into the gas supply 7) _____ after regulating pressure.

城市燃气专业英语 07

Part 2 Speaking

1. Useful sentences

(1) Storage and distribution station combines gas storage station with gas distribution station.

(2) According to the pressure, storage and distribution station is divided into the high-pressure storage and distribution station and the low-pressure storage and distribution station.

(3) The main facility of gas storage station is **gasholder**.

- 37 -

(4) **Gate Station** is the final gas distribution station in the long-distance pipeline.

(5) Before **pressure test** of gas pipeline, it should be pigged no less than twice with a pig.

(6) The non-interruption airtight **pigging** process is adopted.

(7) There are two kinds of **dust collecting** equipment: **cyclone** dust collector and **filter**.

2. Oral practice
Dialogue

A: Good morning! Zhang Ming. Nice to meet you!

B: Morning, Xiao Li! Nice to meet you, too! I heard you attended the lecture about the transmission and distribution system of city gas.

A: Yes, very impressive. I have learned a lot through the lecture.

B: Great! I am writing a paper about it recently. May I ask some questions?

A: OK, no problem.

B: In the gas storage and distribution station, there is a process called odorization, what is its purpose?

A: In order to use natural gas safely, once natural gas leaks out from the pipe and equipment, odorization makes people sense immediately, and in the meantime, it is also a method of leak detection of pipeline.

B: At present, the commonly used odorant in domestic and foreign cities and towns gas is **tetrahydrothiophene (THT)**.

A: Right. In general, 16 ~ 32g THT is added to natural gas of per 1000 cubic meters.

B: Ok, I know! Thanks a lot. I learned really important knowledge from you today.

A: It's my pleasure. See you tomorrow!

B: See you!

Part 3   Reading

Gate Station Process

Gate station is responsible for receiving gas source (including coal gas preparing plant, natural gas, coal bed gas and the plant having residual gas to be used, and so on, which can be used in the town), making the metering and quality test of gas, and according to the requirement of transmission and distribution of gas supply in the city, controlling and adjusting the flow and pressure of the gas which is supplied to the city, and in case of necessity, also making **purification** and odorization of gas.

The selection of gate station site:

1. Fire prevention distance between gate station and surrounding buildings must accord with the regulation of current national standard, which is "Code for Fire Prevention Design of Buildings";

Module 3 Process in Gas Transmission and Distribution

2. Site of gate station should have some conditions, such as the suitable terrain, engineering geology, power supply, water supply and **drainage** and communication, etc.;

3. Site of gate station should be close to the central region of city gas load.

Process of gate station: natural gas is transported to total inlet valve of gate station from high-pressure pipeline, enters the portal tract and filtered and measured by three ways (two open and one standby), and then through pressure regulating system in parallel of two ways (one open and one standby), and after pressure being regulated from 6.0 to 4.0MPa, flows into the underground gas pipeline by the odorization device and gets into the gas station.

Part 4 Translating

(1) 储罐区分类：(1) 可燃液体地上储罐区（常压罐区）；(2) 液化烃、可燃气体、助燃气体地上储罐区（压力罐区）。

(2) 常压储罐是设计压力小于或等于 6.9kPa（罐顶表压）的储罐。

(3) 可能散发可燃气体的工艺装置、罐组、装卸区或全厂性污水处理厂等设施宜布置在人员集中场所及明火或散发火花地点的全年最小频率风向的上风侧。

(4) 液化烃罐组或可燃液体罐组不宜紧靠排洪沟布置。

(5) 低压储罐是设计压力大于 6.9kPa 且小于 0.1MPa（罐顶表压）的储罐。

(6) **Pressure tank** of natural gas is the one whose design pressure is not less than 0.1MPa (gauge pressure of tank top).

(7) Foam station of tank area should be arranged in the non-explosion area out of fire **dike** in the tank field, and its fire protection distance from flammable liquid tank is not less than 20 meters.

(8) It is not suitable to plant **hedgerow** or dense bushes between the processing units or tank field of flammable gas, liquefied hydrocarbon, flammable liquid and **peripheral** fire lane.

(9) People can plant the evergreen **turf** containing more water, whose height is no more than 15 centimeters in the fire dike of tank field of flammable liquid.

(10) It is forbidden to green in the fire dike of tank field of liquefied hydrocarbon.

Part 5 Vocabulary

distribution[ˌdɪstrɪˈbjuːʃn]　　　　　　n. 分配；分布
aeration[eəˈreɪʃn]　　　　　　　　　　n. 通气；充气
compressor[kəmˈpresə(r)]　　　　　　n. 压缩机
gasholder[ˈɡæshəʊldə(r)]　　　　　　 n. 储气罐
pigging[ˈpɪɡɪŋ]　　　　　　　　　　　n. 清管；清管作业
cyclone[ˈsaɪkləʊn]　　　　　　　　　 n. 旋风除尘器
filter[ˈfɪltə(r)]　　　　　　　　　　　 n. 过滤器
　　　　　　　　　　　　　　　　　　vt. 过滤

tetrahydrothiophene[tetrəhaɪdrəˈθaɪəfiːn]	n. 四氢噻吩 (THT)
purification[ˌpjuərɪfɪˈkeɪʃn]	n. 净化
drainage[ˈdreɪnɪdʒ]	n. 排水
flammable[ˈflæməbl]	adj. 易燃的；可燃的
dike[daɪk]	n. 堤
hedgerow[ˈhedʒru]	n. 绿篱
peripheral[pəˈrɪfərəl]	adj. 周围的；外围的
turf[təːf]	n. 草坪；草皮
hydrocarbon[ˌhaɪdrəˈkɑːbən]	n. 碳氢化合物；烃
natural gas storage and distribution station	天然气储配站
gate station	门站
pressure test	试压
dust collecting	除尘
atmospheric tank	常压罐
liquefied hydrocarbon	液化烃
pressure tank	压力罐
combustion-supporting gas	助燃气体

Module 4

Equipment and Facilities

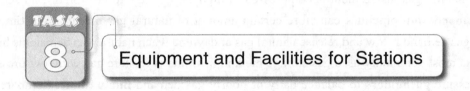

Part 1 Listening

Listen to the following passage and fill in the blanks with proper words.

Gas storage facilities

Gas demand of a city is not constant and varies with time. But gas source supply is almost constant and does not change with time. To solve the **confliction** of constant supply and inconstant demand, city gas distribution system is equipped with gas 1) _____ station. Gas storage station stores gas during low demand and releases gas at 2) _____ demand. The following are several commonly used storage methods. At low gas demand season, gas is 3) _____ into underground space with proper geological **structure**. At high gas demand season, gas is withdrawn from the underground space. The proper geological structures are usually **exhausted** oil or gas fields, underground structure containing water and porosities, salty ore beds and caves, etc.

The most economical method is to store city gas in exhausted oil or gas fields. Underground storage reservoirs can store huge volume of gas with small investment and operation costs. This method saves thousand tons of 4) _____ materials. It is usually used for seasonal peak-shaving and partial daily peak-shaving. It is an ideal storage method for city gas. The difficult part is how to find an economical and suitable geological structure near the city to store gas. The capacity of the first underground storage reservoir in China is 10 million cubic meters. Large scale underground storage reservoirs have been under construction since the beginning of Western to Eastern Project. The volume of liquefied natural gas is much smaller, 1/600 volume of natural gas of the same mass. It can be stored in a 5) _____ insulated holder. Liquefied natural gas will be gasified at gas demanding peak hours and the gas will be supplied to users.

Now LNG has become an international **commodity** because it is easy to be transported and traded between countries. Recently, global LNG production and trade are getting 6) _____. LNG is booming in oil and gas industry.

Gas storage by high pressure pipelines and end section of long transmission pipelines is an effective way of hourly peak-shaving. Groups of high pressure pipelines are buried underground and the natural gas inside them is compressed to high 7) _____ to be stored. End section of long **transmission** pipelines can store certain amount of natural gas during night time when natural gas demand is low and release natural gas at daytime when natural gas demand is high.

The most commonly used gas storage method in China is to store gas with low-pressure or high-pressure **gasholder**s to balance daily or hourly gas demand **fluctuations**. Compared with other storage methods, gasholder storage consumes more metal materials and costs more. But when other gas storage methods are not available, gasholder storage is the only way to store gas.

Part 2 Speaking

1. Useful sentences

(1) Natural gas is usually stored in a liquefied state at a temperature as low as -162 ℃ in order to decrease its volume and facilitate transportation.

(2) Compressed natural gas is typically stored at pressure up to 24.821MPa in cylindrical steel tanks.

(3) Fuel gas contains less carbon than coal and oil. Thus, its combustion **emits** less CO_2 (carbon dioxide) than coal and oil do.

(4) Fuel gas is easy to ignite and its combustion is easy to control with high combustion efficiency.

(5) Natural gas contains the least carbon among coal gas and liquefied petroleum gas and therefore, is called the green fuel.

(6) Development of city gas supply is an important part of China's energy strategies in the

21st century.

(7) The distribution system of city gas is responsible to safely transport gas from sources to residential, commercial and industrial consumers.

(8) The distribution system of city gas is the key part to supply gas to thousands of consumers in a city.

(9) The primary function of any gas **regulator** is to match the flow of gas through the regulator to the demand for gas placed upon the system.

(10) The regulator must maintain the system pressure within certain acceptable limits.

(11) There are three basic operating components in most regulators: a loading mechanism, a sensing element and a control element.

(12) By definition, compressors are intended to compress a substance in a gaseous state.

(13) Once the suction and discharge pressures, the suction gas temperature, the required flow rate and the gas composition are determined, a compressor can be selected to do the job.

(14) A centrifugal compressor is a "dynamic" machine. It can produce a continuous flow of fluid which receives energy from integral shaft **impellers**.

(15) Natural gas enters the compressor station and then goes through the dryer to the compressor.

(16) LNG Stations vary in expenditure depending on the amount of fuel dispensed per hour and the amount of storage on site.

2. Pair work

In this part, you are expected to read aloud all the expressions provided and then make up a dialogue with your partner combined with what you've learned in this section. You can choose any of the topics given below:

——the types of gas storage facilities.
——the use of natural gas.
——the use of gas regulators.
——the use of compressors.

Part 3　Reading

Pressure Regulators

Pressure regulators are found in many common home and industrial applications. For example, pressure regulators are used in gas grills to regulate **propane** pressure, in home **furnace**s to regulate natural gas, in medical/dental equipment to regulate oxygen and **anesthesia** gases, in **pneumatic** automation systems to regulate compressed air, in engines to regulate fuel pressure and in fuel cells to regulate hydrogen pressure. Although the applications vary considerably, the pressure regulators provide the same function. Pressure regulators reduce a supply pressure to a lower outlet pressure and they maintain this outlet pressure regardless of inlet pressure fluctuations. This reduction in pressure is the key characteristic of pressure regulators. The outlet pressure is always at a pressure below the inlet pressure.

When choosing a pressure regulator, many factors need considering. Important considerations include material selection, operating pressures (inlet and outlet), flow requirements, fluid used (gas, liquid, **hazardous** or inert), temperature and so on.

Material Selection

A wide range of materials are available including but not limited to corrosion resistant stainless steels, **brass**, **aluminum** and plastic. Stainless steel provides long life and is ideal for clean rooms and corrosive fluids. When cost is a major contributing factor, brass, aluminum and plastic regulators may be the best option. Aluminum is light weight while plastic is suitable for many medical applications involving bodily fluids. Plastic products are often ideal when a throw-away item is required.

Operating Pressures

The inlet and outlet pressures are important factors to consider before choosing the best regulator. Important questions to answer are: What is the range of fluctuation in the inlet pressure? What is the required outlet pressure? What is the allowable variation in outlet pressure?

Flow Requirements

What is the maximum flow rate that the application requires? How much does the flow rate vary? Porting requirements are also an important consideration.

Fluid-used Gas, Liquid, Hazardous or Inert

It is also important to consider the chemical properties of the fluid before determining the best materials for your application. Each fluid will have its own unique characteristics, so care must be taken to select the proper materials that will come in contact with the process fluid. It is also important to determine if the fluid is **flammable** or hazardous in nature. A non-relieving regulator is preferred for use with hazardous, explosive or expensive gases since no gas is allowed to vent to atmosphere. A relieving regulator will vent excess downstream pressure to atmosphere. The excess fluid should be vented safely in accordance to all safety regulations.

Temperature

The materials selected for the pressure regulator not only need to be compatible with the fluid but also must be able to function properly at the expected operating temperature. The primary **concern** is whether or not the chosen **elastomer** will function properly throughout the expected temperature range. Additionally, the operating temperature may affect flow capacity and/or the spring rate in extreme applications.

Pressure Regulators in Operation

A pressure regulator is comprised of three functional elements: a pressure reducing or restrictive element (generally a **poppet** valve), a sensing element (generally a **diaphragm** or **piston**) and a reference force element (generally a spring). In operation, the spring produces a force which opens the valve. Pressure introduced into the inlet port then flows through the valve and then presses against the sensing device (diaphragm or piston). The regulated pressure acts on the sensing element to produce a force which opposes the spring force and closes the valve.

Part 4   Translating

Compressors

Compressors become essential because gas transmission pipelines extend great distances from the gas field. When natural gas does not have sufficient potential energy to flow, a compressor station is needed. Five types of compressor stations are generally utilized in the natural gas production industry:

(1) Field gas-gathering stations gather gas from wells in which pressure is **insufficient** to produce at a desired rate of flow into a transmission or distribution system. These stations generally handle suction pressures from below atmospheric pressure to 5.17MPa and volumes from a few thousand to many million cubic feet per day.

(2) Relay or main line stations are used to boost pressure in transmission lines. They generally compress large volumes of gas at a pressure range between 1.38MPa and 8.96MPa.

(3) Re-pressuring or recycling stations provide gas pressures as high as 41.37MPa for processing or secondary oil recovery projects.

(4) Field storage stations are used to compress trunk line gas for injection into storage wells at pressures up to 27.58MPa.

(5) Gas distributing stations are used to pump gas from gas source to medium-pressure or high-pressure distribution lines at about 0.14 to 0.69MPa, or pump into bottle storage up to 17.24MPa.

Types of Compressors

The compressors used in today's natural gas production industry fall into two distinct types: **reciprocating** and **rotary** compressors.

Reciprocating Compressors

Reciprocating compressors are most commonly used in the natural gas industry. They are built for practically all pressures and **volumetric** capacities. Reciprocating compressors have more moving parts and, therefore, lower mechanical efficiencies than rotary compressors. Each **cylinder** assembly of a reciprocating compressor consists of a **piston**, cylinder, cylinder heads, suction and discharge valves, and other necessary parts to convert **rotary** motion to reciprocation motion.

A reciprocating compressor is designed for a certain range of **compression** ratios through the selection of proper piston displacement and clearance volume within the cylinder. This clearance volume can either be fixed or variable, depending on the extent of the operation range and the percent of load variation desired.

A typical reciprocating compressor can deliver a **volumetric** gas flow rate up to 30,000 cubic feet per minute (cfm) at a discharge pressure up to 68.95MPa.

Rotary Compressors

Rotary compressors are divided into two classes: the centrifugal compressor and the **rotary** blower.

A centrifugal compressor consists of a housing with flow passages, a rotating shaft on which the impeller is mounted, bearings and seals to prevent gas from escaping along the shaft. Centrifugal compressors have few moving parts because only the impeller and shaft rotate. Thus, its efficiency is high and **lubrication** oil consumption and maintenance costs are low. Cooling water is normally unnecessary because of lower compression ratio and lower friction loss. Compression rates of centrifugal compressors are lower because of the absence of positive displacement. Centrifugal compressors compress gas using centrifugal force. In this type of compressor, work is done on the gas by an impeller. Gas is then discharged at a high **velocity** into a **diffuser** where the velocity is reduced and its kinetic energy is converted to static pressure. Unlike reciprocating compressors, all this is done without **confinement** and physical squeezing.

Centrifugal compressors with relatively unrestricted passages and continuous flow are inherently high-capacity, low-pressure ratio machines that adapt easily to series arrangements within a station. In this way, each compressor is only required to meet compression ratio in station. Typically, its volume is more than 100,000 cfm and discharge pressure is up to 0.69MPa.

A rotary blower is built of a casing in which one or more impellers rotate in opposite directions. Rotary blowers are primarily used in distribution systems where the pressure difference between suction and discharge is less than 0.103MPa. They are also used for refrigeration and regeneration. The rotary blower has several advantages: large quantities of low-pressure gas can be handled at comparatively low **horsepower**; low initial cost and maintenance charge; simple to install and easy to operate and maintain; it occupies minimum floor space, and there is little pulsating flow. Its disadvantages are that it cannot withstand high pressures, it has noisy operation due to gear noise and clattering impellers, improper seal between the impellers and the casing, and it overheats if operated above safe pressures. Typically, rotary blowers can deliver a **volumetric** gas flow up to 17,000cfm, and have a maximum suction pressure of 0.07MPa and a pressure difference of 0.07MPa.

When selecting a compressor, the pressure-volume characteristics and the type of driver must be considered. Small rotary compressors (vane or impeller type) are generally driven by electric motors. Large-volume positive compressors operate at lower speeds and are usually driven by steam or gas engines. They may be driven through reduction gearing by steam turbines or an electric motor. **Reciprocating** compressors driven by steam turbines or electric motors are most widely used in the natural gas industry as the conventional high-speed compression machine. Selection of compressors requires considerations of **volumetric** gas deliverability, pressure, compression ratio and horsepower and so on.

Module 4 Equipment and Facilities

Part 5　Vocabulary

confliction[kənˈflɪkʃən;ˈkɔnflɪkʃən]	n. 冲突，抵触；矛盾，对立，分歧
structure[ˈstrʌktʃə]	n. 结构；构造；建筑物
exhaust[ɪgˈzɔːst]	vt. 排出，耗尽
	vi. 排气
commodity[kəˈmɔdɪtɪ]	n. 商品，货物；农产品；矿产品
transmission[trænzˈmɪʃən]	n. 传送；传递；传导；传达
gasholder[ˈgæshəuldə]	n. 煤气库，气柜，储气罐
fluctuation[flʌktʃuˈeɪʃ(ə)n]	n. 起伏，波动
emit[ɪˈmɪt]	vt. 发出，放射；发行；发表
regulator[ˈregjuleɪtə]	n. 调节器，稳流器，调节阀
impeller[ɪmˈpelə]	n. 叶轮，叶轮片；转子
propane[ˈprəupeɪn]	n. 丙烷
furnace[ˈfəːnɪs]	n. 炉子，火炉；熔炉
anesthesia[ænɪsˈθiːzɪə]	n. 麻醉；麻木
pneumatic[njuːˈmætɪk]	adj. 气动的；充气的
hazardous[ˈhæzədəs]	adj. 有危险的；冒险的
brass[brɑːs]	n. 黄铜；黄铜制品
aluminum[əˈluːmɪnəm]	n. 铝
flammable[ˈflæməb(ə)l]	adj. 易燃的；可燃的；可燃性的
elastomer[ɪˈlæstəmə]	n. 弹性体，高弹体，（高）弹性塑料
concern[kənˈsɜːn]	vt. 涉及，关系到；使担心
	n. 关系；关心；关心的事
poppet[ˈpɔpɪt]	n. 提升阀，盘形活门
diaphragm[ˈdaɪəfræm]	n. 膈；隔膜，隔板
piston[ˈpɪst(ə)n]	n. 活塞
insufficient[ɪnsəˈfɪʃ(ə)nt]	adj. 不足的，不充足的
reciprocating[rɪˈsɪprə,keɪtɪŋ]	adj. 往复的；交互的
rotary[ˈrəutərɪ]	adj. 旋转的；转动的
cylinder[ˈsɪlɪndə]	n. 圆筒；气缸
compression[kəmˈpreʃən]	n. 压缩，浓缩；压榨，压迫
volumetric[vɔljuˈmetrɪk]	adj. 容量的，容积的；体积的
lubrication[luːbrɪˈkeɪʃən]	n. 润滑；润滑作用
velocity[vəˈlɔsətɪ]	n. 速率，速度
diffuser[dɪˈfjuːzə]	n. 扩压段，扩压器
confinement[kənˈfaɪnmənt]	n. 局限，限制；界限；约束
horsepower[ˈhɔːspauə]	n. 马力（功率单位），功率

TASK 9 Accessory Equipment for Gas Pipeline

Part 1 Listening

Listen to the following passage and fill in the blanks with proper words.

Water Drainers

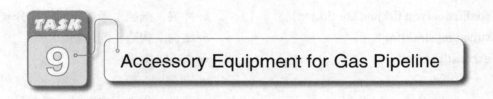

Water drainers are used to 1) _____ the **condensed** water continuously in the pipeline network in the case of normal water seal height on which gas can't be discharged. Therefore, water drainers are necessary safety **affiliated** facilities. Usually 2) _____ pipeline drainers can be built into **vertical** and **horizontal** drainers. In order to keep the 3) _____ height of drainers and easy operation and maintenance, there are usually single water seal and double water 4) _____ forms.

Gate valves should be installed for the drainers of gas pipelines, and be vertical connection with a collection funnel drain, and a drainer is 5) _____, but that will increase partial 6) _____ for the pipeline when its foundation sinks. Drainers can be located outdoors, but anti-freezing measures must be taken in cold areas. When they are located indoors, there should be a good natural 7) _____.

Part 2 Speaking

1. Useful sentences

(1) Steam traps are installed wherever steam is used and they are automatic valves.

(2) The basic function of steam traps is to allow condensate to flow, while preventing the passage of steam until it has given up its heat by condensing back to water.

(3) A valve is a mechanical device that controls the flow of fluid and the pressure within a system or process.

(4) A multitude of valve types and designs safely accommodate a wide variety of industrial applications.

(5) The body, sometimes called the shell, is the primary pressure boundary of a valve.

(6) Valve bodies are usually cast or forged into a variety of shapes.

(7) The cover for the opening in the valve body is called the bonnet.

(8) The internal elements of a valve are collectively referred to as a valve's trim.

(9) For a valve having a bonnet, the disk is the third primary pressure boundary.

(10) The seat or seal rings provide the seating surface for the disk.

(11) The stem, which connects the actuator and disk, is responsible for positioning the disk.

(12) The actuator operates the stem and disk assembly.

(13) An actuator may be a manually operated **handwheel**, manual lever, motor operator, solenoid operator, pneumatic operator or hydraulic ram.

(14) Most valves use some form of packing to prevent leakage from the space between the stem and the bonnet.

(15) Because of the diversity of the types of systems, fluids and environments in which valves must operate, a vast array of valve types have been developed.

2. Pair work

In this part, you are expected to read aloud all the expressions provided and then make up a dialogue with your partner combined with what you've learned in this section. You can choose any of the topics given below:

——the use of valves.

——the composition of valves.

——the importance of valves.

——the types of valves.

Part 3   Reading

Valves

A valve is a device for isolating or regulating the flowrate of gases, liquids and slurries through pipework and launder systems. The force required to operate a valve can be carried out either manually (by hand) or mechanically. Mechanical **attachments** (actuators) to a valve are usually either electrically or pneumatically operated. The actuators can be controlled manually (ie. a technician pushing a button or switch) or by the plant control system. There are hundreds of different types of valves available.

Ball Valves

Ball valves, as the name implies, have a ball with a hole drilled through the centre swivel mounted within the valve body. When the hole in the ball is **orientated** in the same direction as the pipe, this will result in full flowrate. As the hole in the ball is oriented away from the direction of the pipe, the flowrate will be restricted and finally cut off completely when the hole is oriented at 90 degrees to the pipe direction. Note that the hole in the ball is a lesser diameter than the nominal bore of the pipe.

Butterfly Valves

Butterfly valves use a similar principle to ball valves. However, instead of a ball mounted in the valve body, it is a circular disc (called a butterfly because the two half circles around the vertical shaft appear like wings). Again the orientation of the butterfly determines the flowrate. When the butterfly is oriented in the same direction as the pipe (i.e. presenting the least cross sectional area to the moving fluid), this will result in full flow. As the butterfly is oriented away from the direction of the pipe, the flowrate will be restricted by the increased area of obstruction to the fluid and finally cut off completely when the butterfly is oriented at 90 degrees to the pipe direction.

Knifegate Valves

Knifegate valves, often just called gate valves, these are used as **isolation** valves. The principle is simply a knife or gate which is dropped in front of the flow. Knifegate valves should never be used in a **restrictive** role (i.e. half open) as the base of the knifegate will wear rapidly and not seal properly when closed. Knifegate valves come in all sizes and can have manual hand wheels or pneumatic actuators to raise and lower the knifegate.

Diaphragm (Saunders) Valves

Diaphragm valves (commonly known as a Saunders, after a popular brand name) work on the principle of a rubber diaphragm or **bladder** opening and closing. Saunders valves are ideal for restrictive or flow control duties (i.e. valve half closed to reduce flowrate). There are two main types of Saunders valves: **weir** type, straight through type. The diaphragm action can be actuated manually or with a pneumatic actuator (the valve body base remains the same).

Check (Non Return) Valves

Check valves or non-return valves are designed to ensure one way flow only. They are usually used in water pipeline systems and installed immediately after the pump. The most common check valve is the disc type (horizontal or vertical). When flow is sufficient, the disc is pushed out. When flow reduces (or reverses if the pump fails), then the disc falls back into a seat blocking the flow.

Pinch Valves

Pinch valves are used for flow control application, usually to regulate another parameter such as **slurry** level or thickener underflow solids. The valve operating principle relies on a flexible section of pipe being flattened between (pinched) two moving bars (like a vice). The tighter the pinch the lower the flowrate. The pinching mechanism can be manual, but it is usually pneumatically operated and controlled by a PLC system. The rubber section will **perish** with time and will need to be periodically replaced. The most common brand of pinch valve is Larox.

Pressure Relief Valves

Obviously as the name suggests, pressure **relief** valves are a safety device designed to open when system pressure (i.e. in a vessel or pipework) becomes too great and may damage equipment or endanger personnel if not relieved. The most common type is the spring operated valve. A feather valve under spring pressure is seated in the valve body and exposed to system pressure. When the system pressure overcomes the spring pressure, the feather valve will move in the seat, creating an exit to atmosphere, allowing the system gas or liquid to escape.

Part 4   Translating

TPR Valves and Discharge Pipes

Temperature/pressure-relief or TPR valves are safety devices installed on water heating appliances, such as boilers and **domestic** water supply heaters. TPRs are designed to automatically release water in the event that pressure or temperature in the water tank exceeds safe levels.

If temperature sensors and safety devices such as TPRs **malfunction**, water in the system may become superheated (exceed the boiling point). Once the tank ruptures and water is exposed to the atmosphere, it will expand into steam almost instantly and occupy approximately 1,600 times its original volume. This process can propel a heating tank like a rocket through multiple floors, causing personal injury and **extensive** property damage.

Water-heating appliance explosions are rare due to the fact that they require a **simultaneous** combination of unusual conditions and failure of **redundant** safety components. These conditions only result from extreme negligence and the use of outdated or **malfunction**ing equipment.

The TPR valve will activate if either water temperature (measured in degrees **Fahrenheit**) or pressure (measured in pounds per square inch [PSI]) exceeds safe levels. The valve should be connected to a discharge pipe (also called a drain line) that runs down the length of the water heater tank. This pipe is responsible for routing hot water released from the TPR to a proper discharge location.

It is critical that discharge pipes meet the following requirements. A discharge pipe should:

(1) be constructed of an approved material, such as CPVC, copper, **polyethylene**, **galvanized** steel, polypropylene or stainless steel. PVC and other non-approved plastics should not be used since they can easily melt.

(2) not be smaller than the diameter of the outlet of the valve it serves (usually no smaller than 3/4in).

(3) not reduce in size from the valve to the air gap (point of discharge).

(4) be as short and straight as possible so as to avoid undue stress on the valve.

(5) be installed so as to drain by flow of **gravity**.

(6) not be trapped, since standing water may become contaminated and backflow into the potable water.

(7) discharge to a floor drain, to an indirect waste receptor or to the outdoors.

(8) not be directly connected to the drainage system to prevent backflow of potentially contaminating the potable water.

(9) discharge through a visible air gap in the same room as the water-heating appliance.

(10) be first piped to an indirect waste receptor such as a bucket through an air gap located in a heated area when discharging to the outdoors in areas subjecting to freezing, since freezing water could block the pipe.

(11) not terminate more than 6 inches (152mm) above the floor or waste receptor.

(12) discharge in a manner that could not cause scalding.

(13) discharge in a manner that could not cause structural or property damage.

(14) discharge to a **termination** point that is readily observable by occupants, because discharge indicates that something is wrong, and to prevent unobserved **termination** capping.

(15) be piped independently of other equipment drains, water heater pans, or **relief** valve discharge piping to the point of discharge.

(16) not have valves anywhere.

(17) not have tee fittings.

(18) not have a threaded connection at the end of the pipe so as to avoid capping.

Leakage and Activation

A properly functioning TPR valve will **eject** a powerful jet of hot water from the discharge pipe when fully activated, not a gentle leak. A leaky TPR valve is an indication that it needs to be replaced. In the rare case that the TPR valve does activate, the homeowner should immediately shut off the water and contact a qualified plumber for assistance and repair.

Inspectors should recommend that homeowners test TPR valves monthly, although inspectors should never do this themselves. The inspector should demonstrate to the homeowner how the main water supply can be shut off, and explain that it can be located at the home's main water supply valve or at the water supply shut-off for the appliance on which the TPR is mounted.

Part 5 Vocabulary

condense[kən'dens]	vi. 浓缩；凝结
	vt. 使浓缩；使压缩
affiliate[ə'fɪlɪeɪt]	vt. 使附属；接纳；使紧密联系
vertical['vɜːtɪk(ə)l]	adj. 垂直的，直立的
horizontal['hɒrɪ'zɒntəl]	adj. 水平的，地平的
ventilation[ventɪ'leɪʃən]	n. 通风，换气，空气流通
handwheel['hændwiːl]	n. 手轮，操纵轮
attachment[ə'tætʃm(ə)nt]	n. 附件；附着物；附属装置
orientate['ɔːrɪənteɪt]	vi. 向东；定向

Module 4 Equipment and Facilities

	vt. 给…定位；使适应
isolation[ˈaɪsəˈleɪʃən]	*n.* 隔离；分离；脱离；孤立
restrictive[rɪˈstrɪktɪv]	*adj.* 限制的；限制性的；约束的
bladder[ˈblædə]	*n.* 囊；泡
weir[wɪə]	*n.* 堰；低坝
slurry[ˈsləːrɪ]	*n.* 悬浮液，泥浆；水泥浆
perish[ˈperɪʃ]	*vt.* 使麻木；毁坏
	vi. 死亡；毁灭；腐烂
relief[rɪˈliːf]	*n.* 减轻；解除；缓解；泄放
domestic[dəuˈmestɪk]	*adj.* 家庭的，家用的
malfunction[mælˈfʌŋ(k)ʃ(ə)n]	*n.* 故障；失灵；疾病
extensive[ɪkˈstensɪv]	*adj.* 广泛的；大量的；广阔的
simultaneous[sɪm(ə)lˈteɪnɪəs]	*adj.* 同时的；同时发生的
redundant[rɪˈdʌndənt]	*adj.* 过多的，过剩的；多余的
Fahrenheit[ˈfær(ə)nhaɪt]	*adj.* 华氏温度计的；华氏的
	n. 华氏温度计；华氏温标
polyethylene[ˌpɒlɪˈeθɪliːn]	*n.* 聚乙烯
galvanized[ˈgælvənaɪzd]	*adj.* 镀锌的，电镀的
gravity[ˈgrævɪtɪ]	*n.* 重力，地心引力
termination[tɜːmɪˈneɪʃ(ə)n]	*n.* 结束，终止
eject[ɪˈdʒekt; ˈiːdʒekt]	*vt.* (从内部)排出；喷射，喷出

TASK 10 Application Equipment for Gas users

Part 1   Listening

Listen to the following passage and fill in the blanks with proper words.

Gas Cookers

Liquefied petroleum gas (LPG) is one of the conventional sources of fuel for cookstoves in the Philippines. The use of LPG as source of fuel is 1) _____ both in the urban and in the rural areas, particularly in places where its supply is readily accessible. The main reasons why LPG is widely adopted for household use are: it is 2) _____ to operate, easy to control, and clean to use because of the blue flame emitted during 3) _____. However, because of the continued increase in the price of oil in the world market, the price of LPG fuel has gone up tremendously and is 4) _____ increasing at a fast rate. At present, an 11kg LPG, which is commonly used by common households for cooking, costs as high as P540 per tank in urban areas or even higher in some places in rural areas. For a 5) _____ household, having four children, one LPG tank can be consumed within 20 to 30 days only depending on the number and 6) _____ of food being cooked. With this problem on the price of LPG fuel, research centers and institutions are challenged to develop a technology for cooking that will utilize 7) _____ sources other than LPG. The potential of biomass as alternative fuel source to replace LPG is a promising option.

Part 2   Speaking

1. Useful sentences

(1) In most countries, household gas from a pipeline is generally **synonymous** with natural gas.

(2) Gas **oven**s were found in most households by the 1920s with top burners and interior ovens.

(3) The **evolution** of gas stoves was delayed until gas lines that could furnish gas to

households became common.

(4) The introduction of the gas stove offered a major relief to the housewife from some of the hard physical work in the home,

(5) Natural gas **vehicles** mainly focus on CNG vehicle and will remain in the near future.

(6) The automotive industries and the whole society have reached a consensus on developing energy-conserving, environmentally-friendly vehicles.

(7) The CNG vehicles currently in use in China are mainly dual-fuel vehicles (CNG, diesel).

(8) Over the last five years, CNG vehicle has developed rapidly.

(9) CNG stations are not enough to supply CNG in China.

(10) Compared with the diesel-powered vehicles, manufacturing costs of natural gas vehicle are higher.

(11) Natural gas vehicle engine requires high reliability for spare parts, because of high temperature.

(12) Natural gas is composed primarily of **methane**, with trace amounts of other **hydrocarbon**s not removed during the refining process.

(13) Natural gas may be especially important for cutting CO_2 emissions from heavy duty vehicles (HDVs), since other options such as electrification appear to be limited.

(14) Vehicle and fuel technology for natural gas is available today and relatively **affordable**.

(15) In general, in terms of pollutant emissions compared to current diesel vehicles, NGVs perform well, particularly in the HDV segment.

2. Pair Work

In this part, you are expected to read aloud all the expressions provided and then make up a dialogue with your partner combined with what you've learned in this section. You can choose any of the topics given below:

——the use of Natural gas vehicles.

——the advantages of Natural gas vehicles.

——the use of gas cookers.

——the use of gas heaters.

Part 3 Reading

Gas Heaters

The first gas heater made use of the same principles of the Bunsen burner invented in the previous year. It was first commercialized by the English company Pettit and Smith in 1856. The flame heats the air locally. This heated air then spreads by **convection**, thus heating the whole room. Today the same principle applies with outdoor patio heaters or "mushroom heaters" which act as giant Bunsen burners. Beginning in 1881 the burner's flame was used to heat a structure made of asbestos, a design patented by Sigismund Leoni, a British engineer. Later, fire clay replaced the asbestos because it is easier to **mould**. Modern gas heaters still work this way although using other refractory material. Modern gas heaters have been further developed to include units which utilize radiant heat technology, rather than the principles of the Bunsen burner. This form of technology does not spread via **convection**, but rather, is absorbed by people and objects in its path. This form of heating is particularly useful for outdoor heating, where it is more economical than heating air that is free to move away.

Flued Heaters

Flued heaters would always be permanently installed. The flue, if properly installed with correct overall height, should extract most heater emissions. A correctly operating flued gas heater is usually safe.

Non-Flued Heaters

Non-flued heaters –also known as unvented heaters, ventfree heaters or flueless fires may be either permanently installed or portable, and sometimes incorporate a catalytic converter. Non-flued heaters can be risky if **appropriate** safety procedures are not followed. There must be **adequate** ventilation – which is a problem due to the ventilation cooling the house, counteracting the heating – they must be kept clean, and they should always be switched off before sleeping. If operating correctly, the main emissions of a non-flued gas heater are water vapour and carbon dioxide. If there is incomplete combustion, toxic products such as carbon **monoxide** and nitrogen dioxide form.

Operation

Home gas heating controls cycle using a mechanical or electronic **thermostat**. Gas flow is actuated with a valve. Ignition is by electric **filament** or pilot light. Flames heat a radiator in the air duct, but outside the flue, convection or a fan may distribute the heat.

Part 4   Translating

Natural Gas Vehicle

A natural gas vehicle (NGV) is an alternative fuel vehicle that uses compressed natural gas (CNG) or liquefied natural gas (LNG) as a cleaner alternative to other fossil fuels. Natural gas vehicles should not be **confused** with vehicles powered by LPG, which is a fuel with a **fundamentally** different composition. Worldwide, there were 14.8 million natural gas vehicles by 2011, led by Iran with 2.86 million, Pakistan (2.85 million), Argentina (2.07 million), Brazil (1.70 million), and India (1.10 million). The Asia-Pacific region leads the world with 6.8 million NGVs, followed by Latin America with 4.2 million vehicles. In the Latin American region almost 90% of NGVs have bi-fuel engines, allowing these vehicles to run on either gasoline or CNG. In Pakistan, almost every vehicle converted to (or manufactured for) alternative fuel use typically retains the capability to run on ordinary gasoline.

As of 2009, the U.S. had a fleet of 114,270 compressed natural gas (CNG) vehicles, mostly buses; 147,030 vehicles running on liquefied petroleum gas (LPG); and 3,176 vehicles liquefied natural gas (LNG). Other countries where natural gas-powered buses are popular include India, Australia, Argentina, and Germany. In OECD countries there are around 500,000 CNG vehicles. Pakistan's market share of NGVs was 61.1% in 2010, followed by Armenia with 32%, and Bolivia with 20%. The number of NGV **refueling** stations has also increased to 18,202 worldwide as of 2010, 10.2% more than the previous year. Existing gasoline-powered vehicles may be converted to run on CNG or LNG, and can be dedicated (running only on natural gas) or bi-fuel (running on either gasoline or natural gas). Diesel engines for heavy trucks and busses can also be converted and can be dedicated with the addition of new heads containing spark ignition systems, or can be run on a blend of diesel and natural gas, with the primary fuel being natural gas and a small amount of diesel fuel being used as an ignition source. An increasing number of vehicles worldwide are being manufactured to run on CNG. Until recently, the Honda Civic GX was the only NGV commercially available in the US market, however now Ford, GM and Ram have bi-fuel offerings in their vehicle lineup. Fords approach is to offer a bi-fuel prep kit as a factory option, and then have the customer choose an authorized partner to install the natural gas equipment. NGV filling stations can be located anywhere that natural gas lines exist. Compressors or liquifaction plants are usually built on large scale, but small home refueling stations where CNG is used are possible. CNG may also be mixed with biogas, produced from **landfills** or wastewater, which doesn't increase the concentration of carbon in the atmosphere. Despite its advantages, the use of natural gas vehicles faces several limitations, including fuel storage and **infrastructure** available for delivery and distribution at fueling stations. CNG must be stored in high pressure cylinders (20.69MPa to 24.82MPa operation pressure), and LNG must be stored in **cryogenic** cylinders (-260°F to -200°F). These cylinders take up more space than gasoline or diesel tanks that can be molded in intricate shapes to store more fuel and use less on-vehicle space. CNG tanks are usually located in the vehicle's trunk or pickup bed, reducing the space available for other cargo. This problem can be solved by installing the tanks under the body of the vehicle, or on the roof, leaving cargo areas free. As with other alternative fuels, other barriers for widespread use of NGVs are natural gas distribution to and at fueling stations as well as the low number of CNG and LNG stations.

Part 5　Vocabulary

synonymous[sɪˈnɒnɪməs]　　　　　　　*adj.* 同义词的；同义的
oven[ˈʌv(ə)n]　　　　　　　　　　　　*n.* 炉，灶；烤炉，烤箱
evolution[ˌiːvəˈluːʃ(ə)n]　　　　　　　　*n.* 演变；进化论；进展
vehicle[ˈviːəkl]　　　　　　　　　　　*n.* 车辆；工具；交通工具

methane['meθeɪn]	n. 甲烷
hydrocarbon[,haɪdrə(u)'ka:b(ə)n]	n. 碳氢化合物
affordable[ə'fɔ:dəbl]	adj. 买得起的，花费得起的
convection[kən'vekʃən]	n. 对流，传送，传递，传导
mould[məuld]	vi. 发霉
appropriate[ə'prəuprɪət]	adj. 适当的；恰当的；合适的
adequate['ædɪkwət]	adj. 充足的；适当的；胜任的
monoxide[mə'nɒksaɪd]	n. 一氧化物
thermostat['θɜ:məstæt]	n. 恒温器；自动调温器
filament['fɪləm(ə)nt]	n. 灯丝；细丝；细线；单纤维
confuse[kən'fju:z]	vt. 使混乱；使困惑
fundamentally[fʌndə'mentəlɪ]	adv. 根本地，从根本上；基础地
refuel[ri:'fjuəl]	vt. 补给燃料
landfill['lænd'fɪl]	n. 垃圾填埋地；垃圾堆
infrastructure['ɪnfrəstrʌktʃə]	n. 基础设施；公共建设
cryogenic[,kraɪə'dʒenɪk]	adj. 冷冻的；低温学的；低温实验法的

Module 5

Gas Network Construction

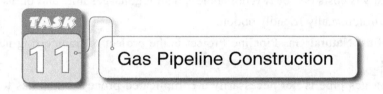

Part1 Listening

Listen to the following passage and fill in the blanks with proper words.

A Pipeline Deal to Exploit America's Fast-Changing Energy Landscape

  Nodding donkeys, offshore 1) _____, refineries and 2) _____ are the bits of the oil industry you can see. A vast and largely 3) _____ network of underground pipes joins them all together. It is worth a lot, which is why Energy Transfer Partners (ETP) said it would pay $5.3 billion for Sunoco on April 30th. It hopes to pull together two networks and shift more of America's booming oil and gas 4) _____.

  ETP is 5) _____. Last year its parent company, Energy Transfer Equity, agreed to buy Southern Union and its gas-pipeline network for $5.7 billion. The latest deal will make ETP the country's second-biggest pipeline firm, behind Kinder Morgan, after the latter's merger with El Paso is concluded later this year.

Sunoco comes with 6) _____, 4,900 filling stations and the remains of a refining business that it is trying to spin off in a 7) _____ with Carlyle, a private-equity firm. But the 8) _____ are the main attraction. ETP currently operates 17,500 miles (28,160km) of the arteries that transport gas and natural-gas liquids such as **propane** and **butane**. Adding Sunoco's 6000 miles, built to carry crude oil and refined products, will reduce ETP's 9) _____ on gas. After the deal, 30% of its 10) _____ will come from oil.

Part2   Speaking

1. Useful sentences

(1) Because of its low density, it is not easy to store natural gas or to transport it by vehicle. Natural gas pipelines are impractical across oceans.

(2) For natural gas, pipelines are constructed of carbon steel and vary in size from 2 to 60 inches (51 to 1, 524mm) in diameter, depending on the different type of pipeline. The gas is **pressurized** by compressor stations and is odorless unless mixed with a **mercaptan odorant**.

(3) Natural gas costs one or two dollars less than regular gasoline and burns cleaner, which makes it an environmentally friendly option.

(4) West-East Natural Gas Pipeline Project is the project of transferring natural gas from west to the east.

(5) Laying gas pipe is not necessarily a complicated process but it has to be completed carefully, because natural gas can leak from pipes that are not securely connected and cause explosions.

2. Oral practice
Dialogue

A: Mr. Li, Are you free now? I have several questions about gas pipelines. I wonder if you could help me with them.

B: No problem. Go ahead.

A: Can gas pipelines be buried in soil?

B: Of course, they can. For example, the gas pipelines used for urban gas network construction are buried. However, in the densely populated area, high-quality steel pipes are required. For example, in the construction of west-east gas pipeline, most of the pipelines are buried in soil. Generally, if long-distance gas pipelines are buried, **cathodic** protection measures must be taken. Nowadays all PE pipelines used for city pipe network are buried. Buried pipelines are usually adopted in cities for the purpose of pleasing appearance.

A: What should be taken into consideration when burying pipelines?

B: It is preferable to slope the pipelines slightly (especially for wet gas) and at the lowest point a **condensate tank** should be set. Protective case pipes are needed while gas pipelines cross roads, banks or railways. Technically, detection pipelines should be set between the casing pipes and the PE pipelines to avoid leakage. Valve chamber and bleeder should be set. Otherwise, the internal air can't be discharged when the pipelines go into operation. If possible, nitrogen filling is an alternative. In that case, you only need to set a valve at one end of the pipeline network, but it is best to use a valve chamber.

A: Which one is more economical, aerial crossing or buried laying?

B: I think it is not a question of which one pays off more, but a question of which one is more feasible. Personally, I think overhead pipelines **corrode** quickly, which significantly increases the maintenance cost. Buried laying is strongly recommended if possible, especially in the construction of municipal medium pressure pipelines. If conditions do not permit, other ways may be chosen.

A: What pipe materials are required for urban gas pipelines?

B: The pipe materials used for transmitting gas must provide enough mechanical strength, have excellent corrosion resistance, shock resistance, air tightness and be easy to connect. The pipe materials which can be chosen are as follows:

(1) Steel pipes have good plasticity, can bear larger stress, and are easy to weld. Depending on different manufacturing methods, steel pipes can be categorized as two types, that is, the seamless steel pipe and the welded steel pipe.

(2) Cast iron pipes, as compared with steel pipes, have excellent corrosion resistance. Therefore, they are widely used in urban low pressure gas pipe network. Cast iron pipes fall into two types: grey and **ductile** cast iron pipe. But cast iron pipes are brittle, not easy to weld and unable to bear large stress, so at important locations and the areas of large dynamic load, steel pipes are needed.

(3) Plastic pipes, rigid **polyvinyl chloride** pipes and **polyethylene** plastic pipes are lighter, and have the advantages of corrosion resistance, low friction resistance, **rigorous interface** and a higher **tensile** strength and simple construction, so they are widely used in the medium and low pressure gas pipes networks. But its aging research remains to be further solved.

A: Thanks for your time, Mr. Li. Your explanations help me a lot.

Part 3   Reading

Natural Gas Pipeline Types

Black metal tube is no longer the only gas transmission material. Many municipal authorities allow construction personnel to install other alternative natural gas pipeline, such as lightweight metal pipeline and plastic ones. Each type of natural gas pipeline tube has its unique advantage, disadvantage and cost. You can select different materials to match your budget and project design if you get command of the materials of the common types of natural gas pipeline.

Corrugated Stainless Steel Tubing

Corrugated stainless steel tubing (CSST) works as a flexible pipe are found in most families which runs through the gas supply lines to electric appliance. Usually CSST are used for entire city supply systems. Both end of the CSST are equipped with pipe fittings through which can be connected to appliances or rigid pipe fittings. Most of the available in the market of CSST has yellow plastic coating or natural metal appearance. The flexibility is its main advantage and CSST can be distorted easily through both obstacles and the wall. The main disadvantage of CSST is its high cost and you should pay more than $15 to buy a 4-foot length of CSST.

High-Density Polyethylene (HDPE)

HDPE is a plastic material that can be used in residential gas pipeline. HDPE gas pipeline are characterized by flexibility which can greatly reduce the installation time, plumbers pull from a roll of a certain length of pipeline and install in the grooves in the interior of the building easily. The traditional iron pipe, must use screw device at both ends while HDPE pipe fittings are connected to both ends by compaction. The base material of HDPE tends to cost more than traditional iron pipe. However, as a plastic material, HDPE is nearly **immune** to **deterioration**, whereas iron or steel will rust eventually.

Black Iron Pipe

The natural gas can be safely wrapped in the thick traditional **black iron** pipeline. An unexpected nail won't easily breakdown black metal pipe unless extreme force. Black iron pipe is the cheapest gas pipeline type because of its popularity and a large supply. However, it is time-consuming to install the black iron pipe though it is easy to find a plumber. Additionally, it tends to rust and corrode when exposed to **moisture**.

Questions

1. Can you list some transmission material for natural gas except black metal pipe?
2. What is HDPE?

Part 4　Translating

(1) Construction of all underground pipelines should be carried out strictly based on design drawings. In the course of **coupling** and connecting fittings, it is necessary to use spacer plate for the space on welding spot.

(2) The pipelines passed heat treatment test should keep away from flames or arcing. During the process of pipelines group welding on the site it is required to use crayon or painting pens to mark map number and seal welding joint serial number for easily performing random nondestructive testing (NDT) and installation.

(3) There is a requirement for workers to report to technicians at once after they find out that there are air vent pipe, draining pipe and piping with slope not indicated in drawings.

(4) For cast iron valves and flange valves of big diameter, there is a danger causing valve flanges damage owing to deformation of flange surface and uneven tightness of bolts, so it is necessary to check flange surfaces to be coupled to make sure there is no distortion resulting from improper welding, prior to the installation.

(5) Special attention should be paid to pipelines to be connected with equipments.

(6) 必须有适当的临时支架支撑（设备），避免由于持久的耐应力、弯曲应力和垂直压力造成设备的故障、过度变形甚至损坏。

(7) 焊缝表面的焊渣必须清理干净，并应进行外观质量检查，看是否有气孔、裂纹、夹杂等焊接缺陷。如存在缺陷必须及时进行返修，并作好返修记录。

(8) 管道支架、吊架的形式、材质、安装位置应正确，数量应达标，牢固程度、焊接质量应合格。

(9) 焊缝及其他应检查的部位应清晰可见。

(10) 焊接工持有劳动部门颁发的《锅炉压力容器焊工考试合格证》方可上岗操作。

Part 5　Vocabulary

exploit[ɪkˈsplɔɪt]	vt. 开发，开拓；剥削；开采
	n. 勋绩；功绩
invisible[ɪnˈvɪzəbl]	adj. 无形的，看不见的；无形的
output[ˈaʊtput]	n. 输出，输出量；产量；出产
	vt. 输出
propane[ˈproʊpen]	n. 丙烷
butane[ˈbjuten]	n. 丁烷
pressurized[ˈpreʃəraɪzd]	adj. 加压的；受压的
	v. 增压；密封；使加压
mercaptan[məˈkæpt(ə)n]	n. 硫醇
odorant[ˈəʊd(ə)r(ə)nt]	n. 有气味的东西
	adj. 有气味的；有香气的
cathodic[ˌkæˈθəʊdɪk]	adj. 阴极的；负极的
corrode[kəˈrəʊd]	vt. 侵蚀；损害
	vi. 受腐蚀；起腐蚀作用
ductile[ˈdʌktaɪl]	adj. 柔软的；易教导的；易延展的
polyvinyl[ˌpɒlɪˈvaɪn(ə)l]	n. 聚乙烯化合物
	adj. 乙烯聚合物的
chloride[ˈklɔraɪd]	n. 氯化物
polyethylene[ˌpɑlɪˈɛθəlin]	n. [高分子] 聚乙烯
rigorous[ˈrɪɡərəs]	adj. 严格的，严厉的；严密的；严酷的
interface[ˈɪntɚˈfes]	n. 界面；接口；交界面

tensile['tensl]	*adj.* 拉力的；可伸长的；可拉长的
high-density['haɪ'densɪtɪ]	*adj.* 高密度的
immune[ɪ'mjun]	*adj.* 免疫的；免于…的，免除的
	n. 免疫者；免除者
deterioration[dɪˌtɪrɪə'reʃən]	*n.* 恶化；退化；堕落
moisture['mɔɪstʃə]	*n.* 水分；湿度；潮湿；降雨量
coupling['kʌplɪŋ]	*n.* 耦合；结合，联结
	v. 连接（couple 的 ing 形式）
polyvinyl chloride	聚氯乙烯
compressor station	压气站
cathodic protection	阴极保护
condensate tank	冷凝槽
black iron	平铁；黑铁板；黑钢板
refinery wastewater	炼油污水；炼油废水
cast iron	铸铁

Module 5 Gas Network Construction

TASK 12 Quality Supervision and Inspection for Gas Pipeline

Part 1 Listening

Listen to the following passage and fill in the blanks with proper words.

The Battle over Gas Prices in Europe

NO one likes getting a gas bill. But Europe's biggest 1) _____ are especially upset over the sums they must pay gas producers, in particular Russia's 2) _____ giant, Gazprom. Some are trying to cut those costs, but with little in the way of leverage over producers their chances of success look 3) _____. Gas prices in continental Europe are mainly set by a decades-old system of long-term contracts, linked to the price of oil. But in 4) _____ liberalised Britain gas is largely traded at spot prices set by current supply and demand. This handed an advantage to some smaller European **utilities** with interconnections to the British spot market when, in 2008, gas prices suddenly fell. Their bigger 5) _____ meanwhile suffered. They were saddled with "take or pay" contracts that 6) _____ them to buy fixed quantities of gas far above what they could sell and at prices way above those on the spot market. The gap between spot and contract prices has not gone away. German firms, which are especially 7) _____ to Russian pipelines, are at a big 8) _____. Yet both Russia and Norway, which supply almost half of Europe's gas, have shown some 9) _____ towards their complaining customers. By the end of 2009 European gas buyers were begging for **relief**, with **oil-indexed** gas then 50% pricier than spot gas. In response, Norway's Statoil allowed an element of spot pricing (around 25% of oil-indexed contracts) for three years. Gazprom responded with a less 10) _____ 15%～20% allowance.

Part 2　Speaking

1. Useful sentences

(1) Permit to work is safety control system to minimize personal injury or damage to property through a written authorization to perform certain work.

(2) Although the permit format and layout may be different, in general there are only three main parts.

(3) On work completion, each permit must be signed off and returned to the point of issue. The work area should be left clean and tidy and all tools, machinery, trash, etc., must be removed from site. It is normal policy to keep permits on file for a period of one year.

(4) Each scaffold shall be designed and constructed so as to ensure that it can withstand the intended load including personnel, materials, tools, lifting equipment and the weight of the scaffold.

(5) Certification of a scaffold is required before it may be used for working purposes.

(6) **Compressed gas cylinders** should be located well away from the actual work area where cutting or welding is in progress.

2. Oral practice

Dialogue

A: Good morning. Are we going to service the pipelines today?

B: Yes. Have you got the tools ready?

A: Yes. Why is the maintenance so frequent?

B: As gas production equipment is working under the conditions of high temperature, high pressure load or **cryogenic** process for long time, it is **liable** to corrosion and wear. So, gas equipment, pipes, valves and instruments will **inevitably** have all kinds of **defects**. If we can't detect the defects in time and take some technical measures to eliminate them, **deformation**, **fracture** and **perforation** may appear, leading to serious fire accidents. To ensure normal production and protect against accidents, we must strengthen detection, maintenance and repair.

A: I see. Are there any special requirements for maintenance personnel?

B: Yes. Maintenance personnel without being trained are not allowed to go on duty. Nobody is allowed to enter the construction site without work permit. All the personnel should receive safety education and pass the examination before the operation.

A: What contents does a work permit involve?

B: It basically includes maintenance items, location, period, maintenance method, potential risk, technical safety measures, maintenance unit and supervisor on-site, security guardian, maintainer, etc.

A: Will we have hot work today?

B: Yes. Please put up the warning sign over there. Do we have enough fire extinguishers?

A: Yes. When are we going to perform combustible gas analysis?

B: We have some preparatory work to do. Firstly we remove the flammable materials near the ignition point.

Part 3   Reading

Acceptance of A City Gate Station

After the construction of the pipelines is finished in the station and the pressure testing for pipe system is qualified, before put into operation, a city gate station will respond to the acceptance. Usually by construction unit, production units and design units to do the on-site process piping for inspection and acceptance. The specific requirements of acceptance are as follows :

(1) The installation process of a city gate station should meet the requirements of the design and production. The process is reasonable, the installation is correct, and the equipment and the pipe are firm and reliable and easy to operate and maintain.

(2) A separate purging inlet must be set when purging, it is not allowed to put dirt into the gas station equipment. Under special circumstances, after the purging, the equipment and valve must be cleaned again.

(3) The installation of the **metering devices** should comply with the measurement procedures.

(4) The installed instruments in the gas station must be by thoroughly validation, and the qualified products must have the factory certificate. Whether on-site installation, or installed on the wall or table board, it must make sure that the instrument are stable, **vibration** phenomenon is not permitted.

(5) Choose a suitable Pressure regulating valves and it must be sensitive, reliable and tight. Manifolds, separators, receivers and launchers and all kinds of valves in the gate station must pass the strength test and **seal test** separately. All the equipment in the gate station must pass the pressure test.

(6) Valves in the gate station must be flexibly operated and be reliable in performance. No rust, leakage and dripping phenomenon can be found. Heating equipment (boiler, **water jacket** furnace) is installed safely and the pressure testing must be qualified. Everything must comply with the design and production requirements when having the trial operation.

(7) The **insulating flanges** work well and with no leakage.

(8) Lightning protection, fire prevention, explosion-proof as well as the environmental protection facilities can be found in the gate station. The following technical data must be provided by the construction unit on the finishing of the process pipeline in the gas station. The design **amendment notification** of **as-built** drawing of process piping construction, **liaison letter**, the inspection and acceptance records of the **concealed work**, pipeline pressure test

records, the summary table of weld inspection, the check and test records of valve pressure and raw material certificate, etc. When the acceptance has been completed, all the above documents can be collected and bound into books by the construction unit, and then be referred to the project owner for file.

Questions
1. What are the specific requirements of acceptance of a city gate station?
2. What should be done when the acceptance has been completed?

Part 4 Translating

(1) An anticorrosive coating performance testing includes cohesive force (peel strength) testing, coating defect detection and insulation resistance of anticorrosive layer measuring.

(2) Cathodic protection effect testing includes soil corrosive survey, protective potential measurement, current measurement and **insulation performance** testing of insulation flanges.

(3) 尽管如此，输气站内埋地金属管道的防腐却一直未能引起足够的重视，一般只采用外加涂层，造成了很严重的腐蚀后果。

(4) 输气站场是整个管输天然气系统的枢纽，输气站中设备安全、高效的运行是保证天然气输送的关键。

(5) 管道系统压力试验之后，气体泄漏性试验之前应进行系统吹扫。

(6) 吹扫压力不得超过容器和管道系统的设计压力。

Part 5 Vocabulary

utilities[juːˈtɪlɪtɪz]	n. 公共事业；实用工具
slender[ˈslendə]	adj. 细长的；苗条的；微薄的
relatively[ˈrelətɪvlɪ]	adv. 相当地；相对地，比较地
disadvantage[dɪsədˈvaːntɪdʒ]	n. 缺点；不利条件；损失
flexibility[ˌfleksɪˈbɪlɪtɪ]	n. 灵活性；弹性；适应性
generous[ˈdʒen(ə)rəs]	adj. 慷慨的，大方的；宽宏大量的；有雅量的
relief[rɪˈliːf]	n. 救济；减轻，解除；安慰；浮雕
liable[ˈlaɪəbl]	adj. 有义务的；应受罚的；有…倾向的；易…的
inevitably[ɪnˈevɪtəblɪ]	adv. 不可避免地；必然地
defect[ˈdɪfekt]	n. 缺点，缺陷；不足之处
deformation[ˌdɪfɔːˈmeʃən]	n. 变形
fracture[ˈfræktʃə]	n. 破裂，断裂；骨折
perforation[ˌpɜːfəˈreʃən]	n. 穿孔；贯穿
as-built[əzˈbɪlt]	adj. 竣工；完工

Module 5 Gas Network Construction

vibration[vaɪˈbreʃən]	n. 振动；犹豫；心灵感应
state-backed	国家支持的
joint venture	合资企业；联合经营
oil-indexed	与石油挂钩的
compressed gas cylinders	压缩气筒
metering device	计量装置
seal test	密封试验
water jacket	水套；冷却管
insulating flanges	绝缘法兰
amendment notification	修改通知
liaison letter	联络单
concealed work	隐蔽工程
insulation performance	绝缘性能

Module 6

Operation and Maintenance

Part 1 Listening

Listen to the following passage and fill in the blanks with proper words.

Operation and Management of City Gate Station

City gate station is the 1) _____ gas distribution station of long-distance gas transmission pipeline, generally 2) _____ at the end of long-distance gas transmission pipeline and the beginning of city gas pipeline network. It is the important facility of city gas transmission and distribution system and also the starting point and **total hub** of city gas transmission and distribution system. The staff of city gate station should strictly 3) _____ the regulations of the security work and fire safety work focused on the vital parts. The staff of city gate station should strictly **implement** the **routing inspection**, duty and **shift system**. At the same time the staff must 4) _____ the various

Module 6 Operation and Maintenance

processes and the testing, installation, operation, maintenance, repair of the equipment used in the 5) _____, and must operate in accordance with the equipment operating 6) _____, maintenance and overhaul procedures and safety technology regulations. City gate station should be equipped with the 7) _____ safety management personnel for operation, maintenance and repair, and should establish the **corresponding** safety target 8) _____ system, safety production management system, safety production responsibility system of post operation and 9) _____ repair plan. The operation and management of city gate station should be 10) _____ with the current national **mandatory** standards.

Part 2 Speaking

1. Useful sentences

(1) The purification devices including the filters, dust collectors and separators are set before the metering, pressure regulating devices and on the mains entering station.

(2) The metering device of city gate station is set up before the pressure regulating device, and is used for gas trade measurement.

(3) The **odorization device** can be set up on the inlet or outlet of city gate station.

(4) The controlled objects of control system are mainly the remotely controlled valves which are set up on the inbound and outbound pipeline.

(5) The operation and management of city gas transmission and distribution system includes gas pipeline and its accessories, city gate stations, gas storage and distribution stations, **regulator station**s, **regulator cabinets (boxes)**, user facilities and gas equipment.

(6) The staff of city gate station on duty should regularly inspect the production area of city gate station according to the specified time.

(7) The staff of city gate station on duty should ensure the normal operation of the metering devices, pressure regulating equipment, pressure vessels, valves, process piping and odorization devices of city gate station.

(8) The process parameters of the production and operation of city gate station should conform to the requirements of the established production process of gas company or be set up according to the directions of the production scheduling department of gas company.

2. Oral practice

Dialogue

A: Good afternoon! Manager Yang.

B: Afternoon! Nice to meet you!

A: I'm afraid I have to take you some time. Please allow me to introduce myself to you, my name is Wang Ming, and I graduated from China University of Petroleum, and I have been working for the gas company for three months. I've got some questions to ask you.

B: Ok, no problem.

A: Can you introduce to me the related knowledge of gas storage and distribution station?

B: Yes. Gas storage and distribution station is the facilities of storage and distribution of natural gas in the city gas transmission and distribution system. Its main task is to carry out the storage, pressure regulating of natural gas, and to distribute gas to the city gas transmission and distribution pipeline network.

A: I see. Can you tell me what a gas storage and distribution station is composed of?

B: A gas storage and distribution station is generally composed of gas storage tank, metering room, substation chamber, power distribution room, control room, pump room, fire pool and other auxiliary facilities of production and life, etc.

A: Ok. Can you briefly explain to me the role of gas storage and distribution station?

B: Of course. First of all, it is to store a certain amount of gas for gas peaking during peak. Then, to ensure a certain degree of gas supply when a temporary fault occurs with the gas transmission facilities or when the pipeline is under maintenance. And then, to mix a variety of used gas to make it uniform in terms of composition. At last, to make gas pressure up or down to ensure an enough pressure for the gas in the transmission and distribution network or before the user appliances.

A: Thanks a lot, I learn a lot from you today.

B: It's my pleasure.

Part 3　Reading

Operation and Management of Gas Storage and Distribution Station

Gas storage and distribution station is the important **infrastructure** of city gas transmission and distribution system. Its main function is to accept the gas supply of gas source and carry on the dust removal, purification, storage, pressure regulating, metering, distribution, quality testing of city gas, and send the gas into the pipeline network of urban or industrial areas after odorization. The main equipments of gas storage and distribution station include the gas storage equipment, **filtration purification devices**, metering devices, pressure regulating devices, measuring instruments, **gas quality testing equipment**, odorization devices, safety devices, **pressurized equipment**, **pigging device**s and **monitoring and control system**, etc.

Gas storage and distribution station is the key part of fire safety, and the staff should strictly enforce the regulations of the security and fire safety. The staff of gas storage and distribution station should strictly implement the **routing inspection**, duty and shift system. At the same time, the staff must be familiar with the various processes and the testing, installation, operation, maintenance, repair of the equipment used in the processes, and must operate in accordance with the equipment operating procedures, maintenance and overhaul procedures and safety technology regulations. The operation of the equipment used in the gas storage and distribution station must strictly implement the safety technology operating regulations without any changes or reducing operations. Overpressure, over-temperature, over-speed and overload are strictly prohibited with the various processes and equipments in the course of storage and transmission of natural gas in gas storage and distribution station. Gas storage and distribution station should be equipped with the full-time safety management personnel for operation, maintenance and repair, and should establish the corresponding safety target responsibility system, safety production management system, safety production responsibility system of post operation and emergency repair plan. According to the principle of "who is in charge, who is in charge of", the master of gas storage and distribution station is responsible for the safety of the station, and the master attendant of production team is responsible for safety management work of the team. The implementation of safety standardization management in the team can ensure the standardization in all aspects like the operating procedures, production operations, production equipment, operating environment, tool placing, protective articles and safety signs. The production and safety facilities of gas storage and distribution station must strictly implement the practice of a synchronous design, synchronous acceptance and synchronous use to ensure the safe production. The operation and management of gas storage and distribution station should be consistent with the current national mandatory standards.

Questions

1. What is the main function of gas storage and distribution station ?

2. What problems must we pay attention to in the operation and management of gas storage and distribution station?

Part4   Translating

(1) 调压装置的巡查内容应包括调压器、过滤器、安全放散设施、仪器、仪表以及其他设备的运行工况，确保不存在泄漏等异常情况。

(2) 高中压调压站一般每季度保养维修一次；中低压调压站每半年保养维修一次。

(3) 常用的调压器按作用原理分类，通常可分为直接作用式调压器和间接作用式调压器。

(4) 调压站验收试压合格后，将燃气通到调压室外总进口阀门处，然后再进行调压站的置换通气。

(5) 调压器的启动应在调压站置换空气合格后进行。

(6) In general, regulator station will go on running once the outlet pressure of regulator station is set.

(7) The regulator is the main equipment in the regulator station, and it is the pressure regulating device which is composed of sensitive components, control components, actuators and valves.

(8) The heating condition of **surge chamber** or **heat insulation** case of a regulator should be checked before the heating period in cold area.

(9) The new or renewable regulator after repair or maintenance cannot be put into operation unless it has gone through **debugging** and met the technology standard.

(10) The regulator station is usually composed of regulators, valves, filters, safety protection devices, bypass pipes and measuring instruments, etc.

Part 5　Vocabulary

implement[ˈɪmplɪm(ə)nt]	*vt.* 实施，执行；使生效，实现；把…填满
	n. 工具，器械；家具；手段
corresponding[ˌkɒrəˈspɒndɪŋ]	*adj.* 相当的，对应的；符合的，一致的
	v. 相符合；类似；相配
mandatory[ˈmændətərɪ]	*adj.* 强制的；命令的；受委托的
infrastructure[ˈɪnfrəstrʌktʃə(r)]	*n.* 基础设施；基础建设
debugging[diːˈbʌgɪŋ]	*n.* 调试
total hub	总枢纽
routing inspection	巡检
shift system	轮班制度
purification device	净化装置
pressure regulating device	调压装置
odorization device	加臭装置
regulator station	调压站
regulator cabinet (box)	调压柜（箱）
filtration purification device	过滤净化装置
gas quality testing equipment	气质检测设备
pressurized equipment	加压设备
pigging device	清管装置
monitoring and control system	监测与控制系统
direct acting regulator	直接作用式调压器
indirect acting regulator	间接作用式调压器
conversion ventilation	置换通气
surge chamber	调压室
heat insulation case	保温情况

Module 6　Operation and Maintenance

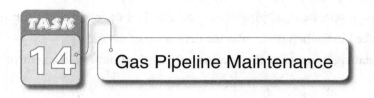

Gas Pipeline Maintenance

Part 1　Listening

Listen to the following passage and fill in the blanks with proper words.

Routine Maintenance of City Gas Pipeline

Routine maintenance of city gas pipeline is the 1) _____ of the normal operation of the pipeline. The main work of routine maintenance is generally divided into two categories: one is to 2) _____ the pipeline itself, 3) _____ the routing inspection (including leak detection, leakage treatment), the **valve chamber maintenance**, pipeline position detection; the other is to 4) _____ the influence of **external** factors on the pipeline, for example the treatment of the 5) _____ slight destruction and damage, illegal buildings. On the city gas pipeline (including **polyethylene pipeline**), the staff 6) _____ carries on the routing inspection, leak inspection and pipeline inspection, 7) _____ is an important work of the routine maintenance of city gas pipeline network. A decision about the 8) _____ of the routing inspection is made according to the region classification, pipe material and operating pressure. 9) _____, the valves, **condensate drainages**, **expansion joints** and other accessories 10) _____ on the city gas pipeline should also be under strict routine maintenance.

Part 2　Speaking

1. Useful sentences

(1) In general, the routing inspection of underground gas pipeline should include the **soil collapse**, landslide, subsidence, earth taking, piling up of garbage or heavy objects, pipe exposure, growing of deep root plants and putting up of buildings, etc., within the scope of the safety protection of gas pipeline.

(2) If the problems are found in the routing inspection, the staff should timely report and take effective treatment measures.

(3) Polyethylene (PE) pipe must not cross the **airtight space**.

(4) The high pressure and sub-high pressure pipeline should be inspected no less than once a year.

(5) The newly operated gas pipe should be checked once within twenty-four hours, and should be rechecked in the first week after the delivering of natural gas.

(6) When damage occurs at the coating of gas pipeline, replacement or repair must be carried out.

(7) When the **blasting engineering** is implemented within the security control area of gas pipeline facilities, the staff should take safety protection measures.

(8) There should be the collision-protection measures and warning signs for the overhead gas pipelines, and the staff should regularly carry out the anti-corrosion inspection and maintenance on the outer surface of gas pipeline.

2. Oral practice

Dialogue

A: Good morning! Mr. Zhang.

B: Morning! Nice to meet you!

A: Are those pigs? Can they clean the pipeline?

B: Yes, no problem. A Pig is a device that moves in the inside of a pipeline for the purpose of cleaning and inspection.

A: Does city gas pipeline also need pigging before being put into use?

B: Yes, for the steel pipe whose **nominal diameter** is not less than one hundred millimeters, pigging should be carried out appropriately.

A: I see. Can you tell me what problems we should pay attention to during pigging operation?

B: Yes, in the course of pigging, the pipeline diameter must be the same specifications, and the pipes of different diameters should be disconnected and cleaned respectively. At the same time, necessary measures should be taken before pigging for the pipe fittings and facilities which will affect the running of pigs.

A: Can you briefly explain to me the qualified requirements of pigging?

B: Hum, after the completion of pigging, when the exhaust is without smoke visually, the staff should install **in the vent** the white cloth or wood board with white paint to inspect if there is any rust, dust or other **debris** within five minutes as qualified. otherwise, the staff can use gas purging to pass.

A: Thanks a lot, I learn a lot from you today.

B: It's my pleasure.

Module 6 Operation and Maintenance

Part 3 Reading

Commissioning of Gas Pipelines

After the installation of gas pipeline is completed, the staff should carry out the **visual inspection**, pigging and purging, **strength test, leak test, completion and acceptance of the project, conversion and production**, etc., in turn.

Visual Inspection

The inspection contents of ground gas pipeline mainly include if the design requirements, visual inspection and **air tightness test** of the pipeline and equipment, and other accessory engineering meet the technical requirements. The inspection contents of underground gas pipeline include the air tightness, tube base, slope, soil cover, depth, **bending angle**, pipe position, operation process, etc.

Pigging and Purging

For **ductile cast iron pipe**, polyethylene pipe, **steel skeleton polyethylene composite pipe** and the steel pipe whose **nominal diameter** is less than one hundred millimeters or whose length is more than one hundred meters, gas purging can be used; for the steel pipe whose nominal diameter is not less than one hundred millimeters, the pigging purging is more suitable. Compressed air can be adopted as the pipeline purging medium, and it is forbidden to use oxygen and **combustible gas**.

Strength Test

In the case of meeting strength test, for the steel pipe whose design pressure is more than 0.8MPa, the clean water should be used to carry on the strength test, otherwise the compressed air is used. The staff should conduct strictly in accordance with the procedures and requirements for the strength test, and should fill in the test record of pipeline system after the qualified strength test.

Leak Test

Leak test should be carried out after the qualified strength test and the backfilling of the whole pipeline trench, except for the checked parts such as the pipe **weld joint**, interface and so on. And these parts are not backfilled until the qualified leak test and anti-corrosion of these parts are both completed. Air is used as the medium for leak test. All equipment, instruments and pipe fittings which did not participate in the leak test should be reset after the qualified leak test, and if there is no leak when the system is **pressurized** according to the design pressure, the test is qualified. The staff should fill in the test record of pipeline system after the qualified leak test.

Completion and Acceptance of the Project

Completion and acceptance of the project is an essential program to inspect the quality of the project, and is also an important measure to ensure the quality of the project. Acceptance procedures must be standardized, rigorous to completely investigate the quality and security risks in engineering.

Conversion and Production

Each group of the conversion and production personnel should be fixed and put into practice in terms of number, posts and responsibilities. Before the conversion, all the **technical disclosure**, safe disclosure and instruction of contingency plan should be carried out with all participants in the conversion according to the conversion program. The convention can be direct with fuel gas, or can be indirect with an **inert gas** as the intermediate medium. Generally large gas transmission and distribution system can use **indirect conversion**, while the small or gas branch pipeline can mostly use **direct conversion**. The staff should conduct strictly in accordance with the procedures and requirements for the conversion. After the conversion is completed, firstly the pressure regulating and safety control devices are adjusted to make the system pressurized to the **predetermined pressure**, and then again a comprehensive inspection will be conducted for the whole system. After confirming in good state, the staff can immediately begin the conversion and gas supply for the users, and after commissioning, the system will be put into normal operation.

Questions

1. What should the staff carry out in turn after the installation of gas pipeline is completed?
2. What problems should we pay attention to in the course of pigging and purging?

Part4　Translating

(1) 对于铁路、道路下面的燃气管道，我们可以通过检查井或检漏管检查是否漏气。

(2) 目前，世界上通用的泄漏检测方法有视觉检漏法、声音检漏法、嗅觉检漏法和示踪剂检漏法。

(3) 对于地面上较高压力燃气管道的泄漏，我们可以迅速关断燃气管网上、下游阀门，以隔断漏气管段。

(4) 当环境浓度在爆炸和中毒浓度范围以内时，工作人员必须进行强制通风，降低浓度后方可继续工作。

(5) 燃气设施泄漏的抢修宜在降低燃气压力或切断气源后进行。

(6) When there is burning at the leakage, measures should firstly be taken to control the fire before reducing pressure or cutting off the gas source, and negative pressure is strictly forbidden.

(7) The rush-repair workers should recheck after the repair and recovering of gas supply, and they can leave the scene of the accident after confirming there are no unsafe factors.

(8) In the rush-repair of liquefied petroleum gas leakage, the effective fire equipment should be provided like **dry powder fire extinguishers**.

(9) In order to prevent water blocking pipe, strict operation management system must be made out to regularly remove the condensed water in the **collecting well**.

(10) The method to remove impurities on the main is to carry out the mechanical cleaning

by segments, commonly with about 50 meters as a cleaning unit.

Part 5　Vocabulary

routine[ruːˈtiːn]	adj. 例行的；常规的；日常的；普通的
external[ɪkˈstɜːnəl;ek-]	adj. 外面的，外部的；外用的；外国的
frequency[ˈfriːkw(ə)nsɪ]	n. 频繁性；频率，次数
debris[ˈdebriː]	n. 碎片，残骸；残渣
pressurized[ˈpreʃəraɪzd]	adj. 增压的；加压的
valve chamber maintenance	阀室维护
polyethylene pipeline	聚乙烯管道
condensate drainage	凝水缸
expansion joint	补偿器
soil collapse	土壤塌陷
blasting engineering	爆破工程
airtight space	密闭空间
nominal diameter	公称直径
in the vent	在排气口
visual inspection	外观检查
strength test	强度试验
leak test	严密性试验
completion and acceptance of the project	工程竣工验收
conversion and production	置换投产
air tightness test	气密性试验
bending angle	弯曲角度
ductile cast iron pipe	球墨铸铁管道
steel skeleton polyethylene composite pipe	钢骨架聚乙烯复合管道
combustible gas	可燃性气体
weld joint	焊缝
technical disclosure	技术交底
inert gas	惰性气体
indirect conversion	间接置换
direct conversion	直接置换
predetermined pressure	预定压力
tracer leak detection method	示踪剂检漏法
rush repair	抢修
dry powder fire extinguisher	干粉灭火器
collecting well	集水井

Module 7

Safety Management

Part 1 Listening

Listen to the following passage and fill in the blanks with proper words.

Type and Function of Gas Stations

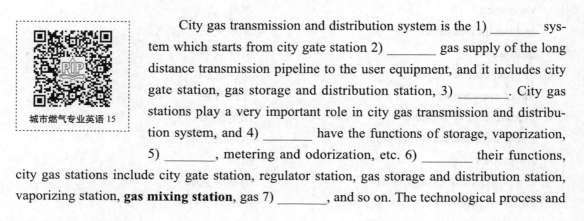

城市燃气专业英语 15

City gas transmission and distribution system is the 1) _____ system which starts from city gate station 2) _____ gas supply of the long distance transmission pipeline to the user equipment, and it includes city gate station, gas storage and distribution station, 3) _____. City gas stations play a very important role in city gas transmission and distribution system, and 4) _____ have the functions of storage, vaporization, 5) _____, metering and odorization, etc. 6) _____ their functions, city gas stations include city gate station, regulator station, gas storage and distribution station, vaporizing station, **gas mixing station**, gas 7) _____, and so on. The technological process and equipment and facilities of city gas stations are also different with their functions 8) _____. We can look on each function as a module, and according to the actual needs, 9) _____ can be made of different modules to achieve the 10) _____ functions.

Part 2 Speaking

1. Useful sentences

(1) Liquefied natural gas **vaporizing station** usually refers to the station with the functions of receiving LNG, storing, vaporizing and transmission.

(2) The processing equipment of liquefied natural gas vaporizing station mainly include tank, vaporizer, pressure regulating device, metering device and low temperature **pump**, etc.

(3) City gate station is not only the terminal gas distribution station of long-distance gas transmission pipeline, but also the receiving station of city natural gas with the functions of purification, pressure regulating, metering, storage and odorization, etc.

(4) The regulator station's task is to adjust the pressure of gas transmission pipeline to the needed pressure of the next level pipeline network or user, and to make the adjusted pressure remain stable.

(5) The station types of liquefied petroleum gas supply system mainly include storage station, **filling station**, storage and distribution station, vaporizing station, gas mixing station and **bottled supply station**, and so on.

(6) Liquefied natural gas industrial system includes natural gas pretreatment, liquefaction, storage, transportation, receiving station and revaporization, and so on.

(7) Natural gas storage and distribution station is the facility for the storage and distribution of natural gas in city gas transmission and distribution system, and its main tasks are gas storage, pressure regulating, as well as the gas distribution to city gas transmission and distribution pipeline network.

(8) Liquefied natural gas vaporizing station mainly serves as gas source of small and medium-sized town that gas transmission pipeline can not reach or that is uneconomic to adopt long-distance gas transmission pipeline, also as gas source of peak shaving and emergency of the town.

2. Oral practice

Dialogue

A: Good morning! Professor Yang, Nice to meet you!

B: Morning! Nice to meet you, too!

A: I'm afraid I have to take you some time. Please allow me to introduce myself to you, my name is Wang Lin, graduated from Northeast Petroleum University, and I have been working in the gas company for only three months. I've got some questions to ask you.

B: Ok, no problem.

A: Can you tell me what the principle of the safety production management of gas stations is?

B: Of course. The principle is to insist on **"safety first, precaution crucial"** in safety production management.

A: Thank you. If there is an explosion or fire accident in gas station, how should we deal with it?

B: Firstly, the on-site staff should immediately notify the control center and the superior leaders. When there are casualties, under the premise of ensuring their own safety, the on-site staff should organize rescue immediately. Secondly, the on-site staff should cut off the upstream gas and the outbound gas quickly and effectively, at the same time they should organize the fire fighting work quickly, and call 119 for aid immediately when the fire is large.

A: Thank you very much indeed. I have learned a lot from you today.

B: It is my pleasure.

Part 3   Reading

Safety Management Regulations for Gas Stations

Safety management for city gas stations involves each process device in station area and pipeline system from multiple aspects, and is the top priority of gas safety production. Therefore, the staff should strictly implement the following safety management regulations:

1. The personnel entering the station must **abide by** the management regulations in station.

2. It is strictly forbidden to smoke and use an open flame in station area.

3. It is strictly prohibited to carry kindling and inflammable and explosive goods into the station.

4. Other vehicles and personnel entering the station must be registered, and should accept inspection initiatively when going out of or into the station.

5. Non-staff (except for visitors) entering the station must have their valid certificates or approval of department manager.

6. Visitors can enter the station with the letter of introduction issued by the superior administrative department, or with the company leaders to accompany.

7. Motor vehicles entering the production area must be added **flash hider**, or enter the station to shut down.

8. It is forbidden to enter the station wearing spiked shoes, and the personnel entering the production area should uniformly wear **anti-static clothing**.

9. The personnel entering the production area of gas stations must turn off mobile phone and other communication equipment.

10. It is forbidden to put to use any fire facilities and tools in station without permission.

11. It is forbidden to take photos and videos in gas stations without the approval of the company.

12. The other business vehicles must be parked at the designated location, and shall not be parked at will.

13. The construction vehicles cannot enter the stations without the construction permit issued by the higher authorities and the fire protective equipment, and must be parked in the specified construction area.

14. Drivers should strictly enforce the regulations for the personnel into the station, and it is forbidden to wander at will in station and to carry irrelevant personnel into the station area.

15. Do well in other security precautions.

Questions

1. Do you know why motor vehicles entering the production area must be added flash hider?

2. Do you know why the personnel entering the production area of gas stations must turn off mobile phones and other communication equipment?

Part 4   Translating

(1) 门站的生产运行必须严格按照公司制定的运行管理制度执行。

(2) 门站的生产运行工艺参数应符合公司制定的生产工艺要求或按照公司生产调度部门的指示进行设定。

(3) 门站的生产设备及工艺管道应按照设备管理的相关规定进行维护保养，保持完好状态。

(4) 门站的压力容器及安全附件应按照国家质量技术监督部门的规定进行定期检定。

(5) 门站工作人员应严格按照操作规程、公司制定的工艺要求和调度指令进行操作。

(6) The staff of city gate station should have the ability to deal with emergency when the production equipment is abnormal.

(7) On-coming person shall have the right to refuse to sign and promptly report to the superior leader when the equipment operation condition that he inspects on the spot is unconformable or unclear with the condition that off-going person provides.

(8) Safety facilities in workplace are all in good condition, and security warning signs are complete and eye-catching.

(9) Post operating personnel cannot take up their posts until they have accepted the security technology training, met the requirements of post security and passed the examination.

(10) The staff of city gate station should regularly maintain fire-fighting equipment and make records.

Part 5   Vocabulary

pump[pʌmp]	n. 泵，抽水机；打气筒
vessel['ves(ə)l]	n. 船，舰；脉管，血管；容器，器皿
gas mixing station	混气站

vaporizing station	气化站
filling station	灌装站
bottled supply station	瓶装供应站
safety first, precaution crucial	安全第一，预防为主
abide by	遵守；信守；承担…的后果
flash hider	防火帽
anti-static clothing	防静电服装
be consistent with	与…一致；符合

Module 7 Safety Management

TASK 16 Safety Management for Gas Pipeline

Part 1 Listening

Listen to the following passage and fill in the blanks with proper words.

Significance of Safety Management for Gas Pipeline Network

The pressure pipeline in city gas transmission and distribution system is a kind of 1) _____ equipment which is widely used in production and life and which may cause the 2) _____ danger of explosion or poisoning accident. The 3) _____ of gas pipeline as a kind of pressure pipelines are that gas pipeline is mostly buried 4) _____, and construction, inspection and maintenance of gas pipeline are difficult through densely 5) _____ areas, limited by 6) _____ conditions and affected by external causes, it is easy for gas pipeline to cause fire, explosion or poisoning accident to 7) _____ the larger social impact and damage. 8) _____, the existing security problems of gas pipeline are analyzed and effective preventive measures are 9) _____ to ensure the secure and stable operation of gas pipeline network, which will have very important practical 10) _____.

城市燃气专业英语 16

Part 2 Speaking

1. Useful sentences

(1) The use safety of gas pipeline can be ensured only in strict accordance with the operating pressure and operating temperature specified in the safety operation procedures of gas pipeline.

(2) Gas pipeline in operation should try to avoid large **fluctuation**s of pressure and temperature, and try to reduce the number of opening and stopping for gas pipeline.

(3) The operator of gas pipeline must go through the learning and training of security technology and post operation, and can't mount guard and operate independently until passing the examination.

(4) The operator of gas pipeline must be familiar with the technical characteristics, system structure, technological process, technological index, possible accidents and the measures which

should be taken of the gas pipeline related to his own post.

(5) The staff of gas pipeline should report timely and deal with the abnormal situation that he found **in the routing inspection**.

(6) The staff of gas pipeline should properly keep all kinds of the records of the routing inspection, in order to prepare for inspection.

(7) In the course of maintenance of gas pipeline, the staff should timely eliminate escaping, emitting, dripping and leaking.

(8) The maintenance work of gas pipeline is the foundation of extending the service life of gas pipeline.

2. Oral practice
Dialogue

A: Teacher Li, Nice to meet you!

B: Zhao Liang! Nice to meet you too!

A: I'm afraid I have to take you some time. I've got some questions to ask you.

B: Ok, no problem.

A: Can you tell me what the methods of safe conversion of natural gas pipeline are?

B: Certainly. There are two kinds of conversion methods of natural gas pipeline: direct and indirect.

A: Thank you. Could you please explain them to me?

B: sure! Firstly, the **direct conversion** method. Process operation of this method is simple, convenient, and more suitable for natural gas pipeline project of some relatively small capacity and low pressure level. As long as the old and new pipelines are connected, we can make the air in the new pipeline converted with the gas pressure, and after the sample test the new pipeline can be put into use. There is a certain risk with this kind of conversion method. Therefore, the **corresponding** safety measures should be adopted. Secondly, the **indirect conversion** method. This method is often used in the new natural gas pipeline engineering of some larger capacity and higher pressure level. First displace the air in the pipeline with an inert gas, and then displace the inert gas with fuel gas. This process is of higher security, but the cost is much higher than the direct conversion method, and the process is complicated.

A: Thank you indeed. I have learned a lot from you today.

B: It is my pleasure.

Part 3   Reading

Archives Management for Gas Pipeline

The technical data of gas pipeline and the management of **archives** are well arranged, which can lay the solid foundation for effective management and use of gas pipeline. In all the work of management of gas pipeline, doing well in the management of technical archives information is of great **significance**. A complete and correct management archive system is established, which can master the quality problem left in the course of the design, manufacture, maintenance, inspection and use of gas pipeline. Depending on the complete and accurate archives information, we can make a reasonable and scientific inspection plan, aiming at the defect inspection, to determine the using conditions and time limit of gas pipeline and further strengthen the management of the whole process of gas pipeline. It can be said that whether the archives information is perfect is one of the important scales to measure the management level of gas pipeline.

The management of the based archives information of gas pipeline should be of high standards, strict requirements, and perseverance, constant attention. Competent departments and enterprises should put forward the principle requirements to the form, specification and content of archives, and the use department must implement in accordance with the unified standards.

The technical archives of gas pipeline mainly include two parts: the original technical data and the use conditions, which constitute all information of the whole process from the design, manufacture, installation, use, inspection and repair of gas pipeline until scrap.

The original technical information refers to the basic information in the course of the design, manufacture and installation of gas pipeline, which is provided by the design, manufacture and construction and installation enterprises. The actual running situation records of gas pipeline include the main process parameters in operation, all previous inspection, repair, transformation and change of gas pipeline, etc., and the shortage, loss or unknown content of pipeline and equipment data caused by historical reasons. All these should be inspected regularly and at least supplement the lack of the necessary information. Each important gas pipeline should have its own archives.

In addition to the technical archives of the various types of pipeline, the archives management for gas pipeline should also need to collect and organize other relevant technical data, including all types of the manuals, **album of paintings**, standards, norms, procedures, systems, major overhaul programs, technology summaries, issued various documents, various report forms and other types of records.

Questions

1. Why must we make archives management for gas pipeline?
2. Do you know what the technical archives of gas pipeline mainly include?

Part4 Translating

(1) 埋地燃气钢质管道检测分为外检测和内检测。

(2) 外检测就是通过非开挖与开挖相结合的方式对管道性能状况进行的检测，包括管道腐蚀环境调查、腐蚀防护状况检测和管体安全状况检测。

(3) 内检测就是通过管道智能爬行器对管道性能状况进行的检测。

(4) 交流电流衰减法可用于管道外防腐层总体状况评价与防腐层破损点定位。

(5) 密间隔电位测试法（CIPS）能够测定防腐层破损面积的大小，并且具有较高的检测准确度，同时可以记录被测管道的阴极保护状态。

(6) **Direct current voltage gradient** (DCVG) technology applies to the buried pipeline with catholic protection, which has higher accuracy of positioning and measurement, and is not affected by the parallel pipeline around.

(7) The principle of **direct current potential method** is to measure **current attenuation** and **potential deviation** through the catholic current to calculate the insulation performance parameters of anti-corrosion insulation layer.

(8) Because city gas pipeline is different from the long distance pipeline, it is one of the important choice projects to use the guided wave detection technology to obtain the information of pipeline body corrosion condition through the non **excavation** method.

(9) When corrosion and other defects appear in the inner wall of city gas pipeline, the detector called pig is commonly used for detection in foreign countries.

(10) Generally, the internal detection device of city gas pipeline is relatively large, expensive, high-cost, and its specificity is strong.

Part 5   Vocabulary

fluctuation[ˌflʌktʃuˈeɪʃ(ə)n;-tju-]	n. 起伏，波动
corresponding[ˌkɒrəˈspɒndɪŋ]	adj. 符合的；相同的；一致的
archives[ˈɑːkaɪvz]	n. 档案，档案室；案卷
significance[sɪgˈnɪfɪk(ə)ns]	n. 意义；重要性；意思
gradient[ˈgreɪdɪənt]	n. 梯度；坡度；倾斜度
	adj. 倾斜的；步行的
excavation[ˌekskəˈveɪʃ(ə)n]	n. 挖掘，发掘
in the routing inspection	在巡检过程中
direct conversion	直接置换
indirect conversion	间接置换
album of painting	图册
external detection	外检测
internal detection	内检测
alternating current (AC) attenuation method	交流电流衰减法
close interval potential survey method (CIPS)	密间隔电位测试法
direct current voltage gradient (DCVG) technology	直流电压梯度技术
direct current potential method	直流电流电位法
current attenuation	电流衰减
potential deviation	电位偏移

Module 7 Safety Management

Task 17 Safety Emergency Response

Part 1 Listening

Listen to the following passage and fill in the blanks with proper words.

Troubleshooting of Gas Pipeline

In the operation of gas pipeline, the 1) _____ problem is the leakage and pipeline 2) _____, and timely discovering leakage is the 3) _____ of the prevention and treatment, and is one of the main tasks of operation management for gas pipeline network. So in the management of gas pipeline, we should 4) _____ catching two problems of the leakage and blockage of gas pipeline, 5) _____ the troubleshooting caused by the leakage and blockage of gas pipeline. The staff should carry out the 6) _____ inspection of pressure pipeline 7) _____ the relevant rules and requirements, and find problems in time and eliminate the hidden dangers to ensure the 8) _____ and safe operation of gas pipeline. 9) _____, the commonly used methods of leak detection in the world include the visual leak detection method, sound leak detection method, the sense of 10) _____ leak detection method and tracer leak detection method.

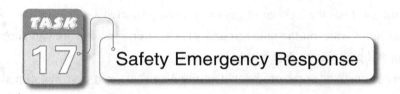

城市燃气专业英语 17

Part 2 Speaking

1. Useful sentences

(1) The **drilling leak detection** is to drill a hole on the ground every certain distance (usually from two to six meters) along the direction of gas pipeline, and to check whether there is a leakage at the hole by smelling or a leak detector.

(2) Test pit digging is to dig pits in the positions of pipeline or pipe joints to reveal the pipes or joints to check if there is a leakage.

(3) Checking the leak by observing the plant growth is a kind of economic and effective method, because gas leaked by the underground pipeline and diffused to the soil will cause the branches and leaves of trees and plants to turn yellow and dry.

(4) If it is found that the pump output of **condensate drainage** increased greatly and suddenly, it is possible that gas pipeline produces cracks, and that groundwater seeps into the condensate drainage. Thus the leakage of gas can also be predicted.

(5) If sand holes leaks on the underground gas cast iron pipeline, we can adopt the method of local drilling and plugging to deal with it.

(6) If there is crack or break on the underground gas cast iron pipeline, **clip sleeve**s can be used to handle it.

(7) In the appearance of leakage point of gas pipeline, especially the higher pressure gas pipeline leakage, the valves on the gas pipeline should quickly be shut off to cut off the leaking pipeline section, and to prevent the accident expanding, and emergency measures should be immediately taken for the treatment of the leakage points.

(8) Due to corrosion perforation, crack, break of steel pipeline and other reasons, when the leakage occurs in the pipeline body, the urgent repair pipeline hoop can be used to rapidly eliminate the accident.

2. Oral practice
Dialogue

A: Good morning! Professor Wang. I have been studying liquefied petroleum gas storage and distribution station by myself, but I meet some difficulties. May I ask you some questions?

B: Of course!

A: Thank you! I saw a question that I am not familiar with on the book. That is about liquefied petroleum gas leak.

B: Hum, liquefied petroleum gas has easy combustion and explosion, easy evaporation, thermal expansion and other dangerous properties, once there is leakage in production, transportation, storage capacity in the process of large storage tank area, it is easy to lead to combustion explosion and poisoning accidents, and even to cause **malignant** environmental pollution accident.

A: I see. Can you introduce some knowledge for the easy-leak location of liquefied petroleum gas?

B: Ok. First, failure or leakage at pipeline **flange**, valve and other connecting sealing parts. Second, the pipeline leakage. Third, leakage at the tank bottom and its valve corrosion. Fourth, tank leakage.

A: What is the treatment technology for the leakage of liquefied petroleum gas (LPG)?

B: Three kinds of treatment technology for LPG leakage are of great significance in the application of LPG tank leakage accidents. They are **leak plugging**, tank transfer and water injection, which can effectively avoid fire, explosion and environment pollution accident. And these technical measures can be popularized and applied to other chemical fuel storage devices.

A: Now I have some ideas about liquefied petroleum gas leak. Thank you very much!

B: You are welcome.

Part 3 Reading

Main Contents of Emergency Plan for Town Gas Accidents

Town gas enterprises should formulate the emergency plan for major safety accident of gas according to the provisions of the relevant laws and regulations of the state and the emergency rescue plan for major safety accident of gas formulated by the local government and administrative department in charge of gas, combined with the specific circumstances of the enterprises; town gas enterprises should perfect the rescue organization institutions, establish professional emergency rescue team, equip with the perfect equipment of rush-repair and rescue, and tools of transportation and communication, and regularly organize training, and actively organize to carry out the training education of the accident emergency rescue knowledge and the publicity work; when the safety accident of gas appears, town gas enterprises should report to the local administrative department in charge of gas in time, and immediately organize the rescue.

According to the functional category in general, emergency plans can be divided into four categories, namely the comprehensive emergency plan, special emergency plan, on-site emergency plan and individual emergency plan. The compiling of the accident emergency plan should be scientific, practical and authoritative. Emergency plan of the rush-repair of city gas facilities should be formulated, and should timely be adjusted and revised according to the specific situation in the safety technical specification for operation, maintenance and rush-repair of city gas facilities. The emergency plans should **be submitted to** relevant departments for the record, and should be exercised on a regular basis no less than once a year.

The emergency plan may include the following main contents:

(1) the basic situation;

(2) dangerous target and its characteristics and the impact on the surrounding;

(3) the available equipment and materials of safety, fire protection, individual protection and their distribution around the dangerous target;

(4) the emergency rescue organization institution, organization personnel and the division of responsibilities;

(5) alarm, communication way;

(6) the treatment measures that should be taken after the accident;

(7) the emergency evacuation and leaving of personnel;

(8) the isolation of hazardous areas;

(9) the detection, rescue, relief and control measures;

(10) the on-the-spot rescue, treatment and hospital treatment of the injured;

(11) the on-the-spot protection;

(12) the emergency rescue support;

(13) plan graded response conditions;

(14) the termination procedure of the accident emergency plan;

(15) the emergency training and the exercise plan of emergency rescue plan.

It should be noted in particular that, once the accident is identified and confirmed, the emergency plan should be started immediately. Only after the following aspects of the completed work can the end of the accident emergency rescue work be confirmed, which is the end of the emergency plan:

(1) the various factors incurring accidents and the dangerous and harmful factors causing accidents have reached the **prescribed** safety conditions, and the production and the living return to normal.

(2) in the process of accident treatment, the water, gas and electricity of being shut down and traffic control to prevent the occurrence of secondary disasters of the accident have resumed normal.

After the end of the accident emergency rescue, by the on-site detection, confirming that the various factors incurring accidents and the dangerous and harmful factors causing accidents have reached the prescribed safety conditions, the accident emergency leading group can order to terminate the instructions of the accident emergency plan, and can notify the relevant departments and local government dangerous release, and the local government can notice the relevant departments on the surrounding and people in the region.

After the actual combat (or exercise) the effect of the emergency plan should be timely evaluated or reviewed, and the emergency plan should be timely **modified** and improved.

Questions

1. How many categories can the emergency plans be divided into according to the functional category? What are they?

2. What contents may the emergency plan mainly include?

Part 4　Translating

(1) 在出口压力过高时，应及时对调压器进行调压，并注意调整后的压力在规定范围内。

(2) 当调压器发生故障时，应及时关闭调压器进口阀，同时使用备用调压器。

(3) 对于直接作用式调压器，由于皮膜破裂，燃气会直接由高压侧流向低压侧，造成高压送气，必须立即更换皮膜。

(4) 对于直接作用式调压器，由于阀口关阀不严或皮膜漏气造成出口压力上升或关闭压力过高，此时应研磨阀口，更换胶垫及皮膜。

(5) 对于直接作用式调压器，由于出口管道有积水，会有明显的出口压力跳动，此时应排除附近管道上的凝水缸的积水。

(6) For **T type regulator**, the rupture of the diaphragm of regulator can often cause the regulator shutting down automatically to stop gas supply. In this case, only the diaphragm needs replacing.

(7) For the **piston type regulator**, the bad installation quality can often cause the fluctuation or beating of the outlet pressure and adjusted pressure. So the technical standard of the regulator assembly must be strictly grasped.

(8) For the **self-operated regulator**, the **diaphragm rupture** of regulator, wear of valve port and damage of the valve pad will cause the shutoff pressure of the regulator to be too high or imprecise. At the moment, the diaphragm and the valve pad should be replaced, and the valve port should be **grinded**.

(9) For the self-operated regulator, the **spring** of the **pilot** failure, the **nozzle** blockage, the valve port blockage of main regulator or the valve pad swell due to corrosion can cause the loss of the outlet pressure. At this time, we should clear the nozzle, clean the regulator, and replace the valve pad and the spring.

(10) For the **meander type regulator**, because the upper valve port of the pilot isn't opened and the outlet **throttle valve** of the pilot is completely blocked, the regulator can't be regulated and started. At this time, we should reassemble the pilot and clean the throttle valve.

Part 5　Vocabulary

malignant[mə'lɪgnənt]	*adj.* 恶性的；有害的；有恶意的
flange[flæn(d)ʒ]	*n.* 法兰；凸缘
prescribe[prɪ'skraɪb]	*vt.* 指定，规定；指定，规定
	vi. 建立规定，法律或指示；开处方
modified['mɔdɪfaɪd]	*adj.* 改进的，修改的；改良的
imprecisely[ˌɪmprɪ'saɪzlɪ]	*adv.* 不严密地
grind[graɪnd]	*vt.* 磨碎；磨快
spring[sprɪŋ]	*n.* 弹簧；春天；泉水；活力；跳跃
pilot['paɪlət]	*n.* 指挥器；飞行员；领航员
nozzle['nɒz(ə)l]	*n.* 喷嘴；管口；鼻
drilling leak detection	钻孔查漏
condensate drainage	凝水缸
clip sleeve	夹子套筒
leak plugging	堵漏
be submitted to	被提交给…；服从于…
rubber gasket	胶垫

T type regulator	T 型调压器
piston type regulator	活塞式调压器
self-operated regulator	自力式调压器
diaphragm (or film) rupture	皮膜破裂
meander type regulator	曲流式调压器
throttle valve	节流阀

Module 8

LNG and CNG

Part 1 Listening

Listen to the following passage and fill in the blanks with proper words.

Falling Oil Prices Affect Nations Differently

World oil prices have dropped by about 30 percent since June. That has helped many economies around the world. When people spend less on 1) _____, they can spend more on other goods. But lower oil prices also mean reduced 2) _____ for national oil companies. Privately-owned oil producers and companies that provide services to the oil industry are also 3) _____. Increased American oil production is one reason for the drop in world oil prices. Technology developed in the U.S. has made that increase possible. **Hydraulic fracturing**, or 4) _____, is a process of pumping water and chemicals into the ground to force out natural gas and oil. It is often used to recover these fossil fuels from rock called 5) _____.

Jim Krane is an energy expert at the James Baker Institute for Public Policy at Rice University. He says oil produced from shale has changed the market.

Slowing economies in China and Europe have also reduced the demand for oil. In the past, the Organization of Petroleum Exporting Countries would react to 6) _____ demand by cutting production. The twelve member states often cooperate to 7) _____ oil prices around the world. However, Jim Krane says members want to defend their share of the market and are unwilling to make production cuts. "They are willing to see prices drop for a little while to see if they could 8) _____ some of these higher-cost producers out of the market." However, this could affect small oil producing companies in the U.S. These companies feel the effects of falling prices more sharply on their profit margin. "If you are a small company thinking about 9) _____ in putting in some new oil production, well, you might think twice if prices are down at $75 (a barrel)." U.S. companies are not alone in feeling pressure from falling prices. Oil makes up 83 percent of Nigeria's exports and 70 percent of the nation's economy. Finance minister Ngozi Okonjo-Iweala recently announced 10) _____ the government would take to increase income.

Part 2 Speaking

1. Useful sentences

(1) Regasification is a process of converting liquefied natural gas (LNG) at −162℃ (−260°F) temperature back to natural gas at atmospheric temperature.

(2) In its liquid state, LNG is not explosive and can not burn.

(3) For LNG to burn, it must vaporize first, then mix with air in the proper proportions (the flammable range is 5 percent to 15 percent), and then be ignited.

(4) In the case of a leak, LNG vaporizes rapidly, turning into a gas, and mixing with air. If this mixture is within the flammable range, there is a risk of ignition which would create fire and thermal radiation hazards.

(5) A typical LNG process is that the gas is first extracted and transported to a processing plant where it is purified by removing condensates such as water, oil, mud, as well as other gases such as CO_2 and H_2S.

2. Oral practice

Dialogue

A: Hello Mr. Smith. Can I ask you several questions about Chinese **oil resources**?

B: Of course. Maybe I'm not able to answer all the questions, but I'm happy to try.

A: Thanks. I'm honored to interview you. What kinds of petrol chemical **feed stock** are there in China?

B: Petroleum, coal and natural gas are major chemical feed stock. Petroleum, as a major energy, and important chemical feed stock, has made great contributions to the development of human society.

A: I can't agree with you more. Could you talk about the petroleum resources in China?

B: No problem! Although China is rich in oil and coal, we have confronted many difficulties.

A: Oh! What are they? Can you tell me in detail?

B: First, due to mass **exploitation**, oil has been facing depletion. Second, its overuse has led to the serious environmental damage and pollution, especially the Green-House effect caused by excess **emission**s of CO_2.

A: Yes, it has become a global problem.

B: At the same time, oil resource is also related to the national economy and the people's livelihood, and is important industry to the development of the country's economy.

A: Facing with the problems, what measures will China take?

B: On one hand, we have to try our best to reduce the pollution. On the other hand, we will take full use of new technology.

A: Oh! I hear that you started to explore new fields for oil and gas resources, and have used new technology, Right?

B: That's right! New areas of oil and gas exploration and the new technology are important objectives during China's 13th 5-year plan period in the energy field and also the focus of today's oil.

A: I see. Thank you for your time Mr. Smith.

Part 3   Reading

Growing Demand for Natural Gas

Natural gas plays an important role both in the U.S. energy supply and in obtaining the nation's economic and environmental goals. The total consumption has begun to exceed **available** domestic natural gas supply though the production in North America is expected to gradually increase up to 2025. As time goes on, the gap between supply and demand will be enlarged.

Increase the import of liquefied natural gas (LNG) to ensure that American consumers get sufficient gas supply in the future is one of the options have been proposed. Liquefaction technology makes this is likely to be "stranded" gas into the main market. Developing countries with abundant natural gas resources are very keen to replace liquefied natural gas (LNG) with money by exporting. On the other hand, developed countries with little or no domestic natural gas resources are mainly relying on imports.

In 1900s, people began to make a lot of efforts to liquefy natural gas for storage, but it wasn't until 1959 that the world's first LNG ship carried LNG from Louisiana to the United Kingdom, which proved the feasibility of overseas liquefied natural gas transportation. Japan first imported LNG from Alaska in 1969 and with a large number of expansion of liquefied natural gas (LNG) imports in the 1970s an 1980s, moved to the forefront of LNG trade. In 1970s, the United

States first imported LNG from Algeria.

Algeria is the world's second largest exporter of liquefied natural gas (LNG).By four liquefaction plants of Algeria's national oil and gas company, Algeria provide liquefied natural gas mainly to Europe (France, Belgium, Spain and Turkey)and the United States. Nigeria mainly exports LNG to Turkey, Italy, France, Portugal and Spain, but also under contract to provide products to the United States in the short term. The republic of Trinidad and Tobago exports liquefied natural gas (LNG) to the United States, Puerto Rico, Spain, and the Dominican republic. An Egyptian factory exports of liquefied natural gas for the first time in 2005, and is expected to supply France, Italy and the United States. Since 2006, Norway plans to export LNG from Melkoya island to the market of Spain, France and the United States. (excerpts from the website)

Questions

1. When the world's first LNG ship carried **cargoes** from Louisiana to the United Kingdom, proving the feasibility of **transoceanic** LNG transport?

2. Which country is the world's second-largest LNG exporter ?

Part 4 Translating

(1) LNG 是液化天然气 liquefied natural gas 的英文缩写。为了方便储存和运输，天然气（主要成分为甲烷）被转化成液态。

(2) 液化天然气无色无味，没有毒性，没有腐蚀性。

(3) 天然气的液化过程包括去除某些组分，比如灰尘、酸性组分、氦、水和重烃。

(4) A：为什么天然气被净化后才能作为燃料使用？

B：未被处理过的天然气含有杂质甚至含有水。杂质被清除后，才能保证天然气持续稳定的燃烧。水会造成管道和设备中水合物的产生，从而引起管道和阀门的阻塞。硫化氢是有毒的气体，同时也是化工产业生产硫磺的原料。二氧化碳会降低天然气的热值，酸性气体和水结合而腐蚀管道和设备。

(5) LNG is transported in specially designed ships with double **hulls** protecting the cargo systems from damage or leaks. There are several special leak test methods available to test the **integrity** of an LNG vessel's membrane cargo tanks.

(6) Transportation and supply is an important aspect of the gas business, since natural gas reserves are normally quite distant from consumer markets. Natural gas has far more volume than oil to transport, and most gas is transported by pipelines.

(7) There is a natural gas pipeline network in the former Soviet Union, Europe and North America. Natural gas is less dense, even at higher pressures. Natural gas will travel much faster than oil through a high-pressure pipeline, but can transmit only about a fifth of the amount of energy per day due to the lower density.

(8) Natural gas is usually liquefied to LNG at the end of the pipeline, prior to shipping.

(9) Short LNG pipelines for use in moving product from LNG vessels to onshore storage are available.

(10) Longer pipelines, which allow vessels to offload LNG at a greater distance from port facilities are under development.

Part 5　Vocabulary

profit[ˈprɑfɪt]	n. 好处，益处；(财产等的)收益
affected[əˈfektɪd]	adj. 受到影响的；做作的；假装的
hydraulic[haɪˈdrɔlɪk]	adj. 液压的；水力的；水力学的
fracture[ˈfræktʃə]	vi. 破裂；折断
decreasing[dɪˈkriːsɪŋ]	adj. 渐减的
	v. 减少，缩减）使降低；降低
exploitation[ˌeksplɔɪˈteɪʃən]	n. 利用，(资源等的)开发，开拓，开采
emission[ɪˈmɪʃən]	n. 发射，散发；喷射；发行
available[əˈveɪləb(ə)l]	adj. 有效的，可得的；可利用的；空闲的
cargo[ˈkɑːgəu]	n. 货物，船货
transoceanic[ˌtrænsəuʃɪˈænɪk]	adj. 横越海洋的；在海洋彼岸的
hulls[hʌlz]	n. 外壳
integrity[ɪnˈtegrɪtɪ]	n. 完整；正直；诚实；廉正
hydraulic fracturing	水力压裂
oil resources	石油资源
feed stock	原料

TASK 19 Compressed Natural Gas

Part 1　Listening

Listen to the following passage and fill in the blanks with proper words.

The Need to Be Seen to Be Clean

　　Natural-gas production is 1) _____, but its green image is in question.

　　There are the signs of America's natural-gas boom. Thanks to new drilling technology, and in particular a 2) _____ process called hydraulic fracturing or "3) _____", the size of the proven reserves is growing. At the end of 2009 the United States had 4) _____ reserves of 283.9 trillion cubic feet (8 trillion cubic meters) of natural gas, up 11% from the year before. In 2010 the country produced 22.6 trillion cubic feet of natural gas, up from 18.9 trillion cubic feet in 2005. The price at the wellhead has dropped from $7.33 per thousand 5) _____ feet to $4.16 during the same period.

　　But some question whether natural gas is really as green as all that. For one thing, fracking uses a 6) _____ amount of water, a severely undervalued resource inland. On the other hand, the process gives off methane, a potent heat-trapper. And fracking raises other concerns. In the process, a mix of sand, water and chemicals is pumped deep underground at 7) _____ pressure. That breaks up 8) _____ formations, releasing the gas trapped inside so it can be pumped to the surface. But fracking is almost entirely unregulated, because of a 2005 **statutory exemption** from the Safe Drinking Water Act. Three members of Congress recently released a report on fracking, saying that oil and gas companies used more than 2,500 fracking products containing 750 chemicals between 2005 and 2009. Some 9) _____ were benign, such as citric acid and instant coffee. Others, though, were extremely toxic, such as 10) _____ and lead.

Module 8　LNG and CNG

Part 2　Speaking

1. Useful sentences

(1) Compressed natural gas (CNG) (Methane stored at high pressure) can be used in place of gasoline (petrol), diesel fuel and **propane**/LPG. CNG **combustion** produces fewer undesirable gases than the fuels mentioned above.

(2) CNG is made by compressing natural gas (which is mainly composed of methane), to less than one percent of the volume it occupies at standard atmospheric pressure. It is stored and **distributed** in hard containers at a pressure of 20 ~ 25MPa, usually in **cylindrical** or **spherical** shapes.

(3) CNG is used in traditional gasoline internal combustion engine cars that have been **converted** into bi-fuel vehicles (gasoline/CNG).

(4) Compressed natural gas vehicles require a greater amount of space for fuel storage than conventional gasoline powered vehicles. Since it is a compressed gas, rather than a liquid like gasoline, CNG takes up more space for each GGE (gasoline gallon equivalent).

(5) Therefore, the tanks used to store the CNG usually take up additional space in the trunk of a car or bed of a pickup truck which runs on CNG.

(6) This problem is solved in factory-built CNG vehicles that install the tanks under the body of the vehicle, leaving the trunk free.

(7) Another option is installation on roof (typical on buses), requiring, however, solution of structural strength issues.

(8) It is safer than other fuels in the event of a spill because it is lighter than air and disperses quickly when released.

2. Oral practice

Dialogue

A: Mr. Li, Can you spare me a few minutes? I have several questions about CNG.

B: Of course, go ahead.

A: What are the pros and cons of using compressed natural gas as an alternative fuel for cars?

B: Natural gas vehicles, or NGVs, make up nearly 120,000 of the cars and trucks that travel on the United States highway systems each year. Compressed natural gas has many advantages over regular gasoline. In comparison to conventional cars running on diesel or gasoline, natural gas vehicles have lower emissions. Owners of NGVs indicate that the vehicles last longer and typically require less maintenance. The fuel economy, acceleration and cruising speed of an NGV typically no less than that of a conventional car.

A: Are gas-powered cars safe?

B: According to the U.S. Department of Energy, overall, NGVs are as safe as gasoline or diesel-fueled vehicles. The fuel tanks in natural gas vehicles are designed to withstand extremes of temperature and heavy **impact**. In the event of a crash, compressed natural gas would evaporate into the air, rather than puddle beneath the vehicle like gasoline or diesel. This could be a safety feature, as the evaporated natural gas is unlikely to be ignited once it has left the tank.

A: Are there any limitations when we choose to drive an NGV?

B: If you want an NGV, you have very few options in the United States. Ford, General Motors and Ram have plans to introduce bi-NGVs someday, which means the vehicles run on both natural gas and regular gasoline. However, Honda already has one NGV in its **lineup**. NGVs also cost more than traditional cars and trucks; in fact, the Honda NGV costs an additional $5,000 for this feature alone. Experts suggest that NGVs have a limited driving **range**, providing 130 miles less than gasoline-powered vehicles.

A: Are there any other disadvantages?

B: The United States has around 1,000 natural **gas fueling stations** across the country, however, the public only has access to about 540. In addition, these fueling stations exist around **metropolitan** hubs and not on the highways. Natural gas refueling devices for your home cost between $2,000 and $5,000, and it takes overnight in order to fuel your vehicle. Once the natural gas reserves run dry, the country has to look for other fuel sources abroad.

Part 3   Reading

The Development of CNG in Asia

In Singapore, more and more public transportation, such as buses, taxis and trucks are increasingly in the use of compressed natural gas.

In 2005, Myanmar's ministry of transportation passed a law requiring all public transport vehicles—cars, trucks and taxis, be converted to compressed natural gas as fuel. The Government allow private companies to do the conversion of the existing diesel and petrol cars, and also began to import CNG powered buses and taxis.

In China, companies such as Sino-Energy are actively in expanding CNG filling stations in the coverage of medium-sized cities in mainland China. There are at least two natural gas pipeline in inland areas are running.

In Pakistan, in 2004, under the command of the Supreme Court, the Karachi government force all city buses and motor tricycle to use CNG in order to reduce air pollution.

In India, the Delhi Transportation Companies operates the fleet of CNG powered buses.

In Pakistan, in 2012, due to the shortage of natural gas, and the negative impact on manufacturing, the federal government has announced plans to gradually stop using CNG within three years.

Iran has the world's largest CNG vehicle fleets and CNG distribution networks. There are

1,800 CNG fueling stations, with a total of 10,352 CNG nozzles. The number of CNG powered vehicles in Iran is about 2.6 million.

Questions

1. What problems that people may face though CNG is increasingly being used by public transport vehicles like buses and taxis, as well as goods vehicles?

2. What is the development of CNG in China?

Part4 Translating

(1) 有一个问题依然存在，那就是如何建立支撑天然气汽车保有量增长的加气站网络。

(2) 目前，意大利是欧洲使用 CNG 汽车最多的国家，也是全世界使用 CNG 汽车数量排名第四的国家。

(3) 我们将探讨压缩天然气加气站的计量、压缩、脱水、储气及高压管道方面的技术问题。

(4) 但在使用天然气作为燃料的过程中，CNG／汽油两用燃料发动机可能出现了一些问题，如发动机磨损、腐蚀，发动机润滑油使用寿命缩短等问题。

(5) 人们经常把 CNG（压缩天然气）与 LNG（液化天然气）相混淆。虽然它们都是天然气存储的形式，但关键的区别是，CNG 是以气体形式在高压下被存储，而液化天然气则是在非常低的温度下存储，且在这一过程中成为液体。

(6) CNG has a lower cost of production and storage compared to LNG as it does not require an expensive cooling process and cryogenic tanks. CNG requires a much larger volume to store the same mass of gasoline or petrol and the use of very high pressures (20.5 to 27.5MPa, or 205 to 275bar).

(7) LNG is often used for transporting natural gas over large distances, in ships, trains or pipelines, and the gas is then converted into CNG before distribution to the end user.

(8) CNG is being **experimentally** stored at lower pressure in a form known as an ANG (adsorbed natural gas) tank at 35 bar (3.5MPa, the pressure of gas in natural gas pipelines), in various sponge-like materials, such as **activated** carbon and MOFs (metal-organic frameworks).

(9) The fuel is stored at similar or greater energy density than CNG. This means that vehicles can be refueled from the natural gas network without extra gas compression; the fuel tanks can be slimmed down and made of lighter, weaker materials.

(10) Compressed natural gas is sometimes mixed with **hydrogen** (HCNG) which increases the H/C ratio (heat capacity ratio) of the fuel and gives it a flame speed about eight times higher than CNG.

Part 5 Vocabulary

controversial[ˌkɒntrəˈvɜːʃl]	adj. 有争议的；有争论的
tremendous[trɪˈmendəs]	adj. 极大的，巨大的；惊人的
ferocious[fəˈrəuʃəs]	adj. 残忍的；惊人的
statutory[stætjut(ə)rɪ]	adj. 法定的；法令的
exemption[ɪgˈzempʃən]	n. 免除，豁免；免税
constraint[kənˈstrent]	n. 约束；态度不自然；强制
propane[ˈprəuˈpen]	n. 丙烷
combustion[kəmˈbʌstʃən]	n. 燃烧，氧化；骚动
distribute[dɪˈstrɪbjuːt;ˈdɪstrɪbjuːt]	vt. 分配；散布；分开
cylindrical[səˈlɪndrɪkl]	adj. 圆柱形的；圆柱体的
spherical[ˈsfɪərɪkəl]	adj. 球形的，球面的；天体的
convert[kənˈvɜːt]	vt. 使转变；转换…
impact[ɪmˈpækt]	n. 影响；效果；碰撞；冲击力
lineup[ˈlaɪnˌʌp]	n. 阵容；一组人；
range[rendʒ]	n. 范围；幅度；排；山脉
metropolitan[ˌmetrəˈpɑlɪtən]	adj. 大都市的；大主教辖区的
predominantly[prɪˈdɑmɪnəntlɪ]	adv. 主要地；显著地
experimentally[ɪkˌsperəˈmentlɪ]	adv. 实验上；用实验方法
activate[ˈæktəˈvet]	vt. 刺激；使活动；使活泼
	vi. 激活；有活力
hydrogen[ˈhaɪdrədʒən]	n. 氢
NGV(Natural Gas Vehicle)	天然气车辆
gas fueling station	加气站
statutory exemption	法定豁免

Vocabulary

A

abide by	遵守；信守；承担……的后果
absolute pressure	绝对压力
acquisition[ˌækwɪˈzɪʃ(ə)n]	n. 获得物，获得
activate[ˈæktəˈveɪt]	vt. 刺激；使活动；使活泼
	vi. 激活；有活力
adequate[ˈædɪkwət]	adj. 充足的；适当的；胜任的
adsorption storage of natural gas(ANG)	天然气的吸附储存
aeration[eəˈreɪʃn]	n. 通气；充气
affected[əˈfektɪd]	adj. 受到影响的；做作的；假装的
affiliate[əˈfɪlieɪt]	vt. 使附属；接纳；使紧密联系
affordable[əˈfɔːdəbl]	adj. 买得起的，花费得起的
air tightness test	气密性试验
airflow[ˈeəfləu]	n. 气流
airtight space	密闭空间
album of painting	图册
allowable pressure drop	允许压降
alternating current (AC) attenuation method	交流电流衰减法
aluminum[əˈluːmɪnəm]	n. 铝

amendment notification	修改通知
ammonia[əˈməunɪə]	n. 氨；氨水
amplitude[ˈæmplɪtjuːd]	n. 振幅；广大，广阔，充足
anesthesia[ænɪsˈθiːzɪə]	n. 麻醉；麻木
annual average daily gas consumption	年平均日用气量
anti-static clothing	防静电服装
appliance connecting pipeline	用具连接管
appropriate[əˈprəuprɪət]	adj. 适当的；恰当的；合适的
aquifer porous formation storage	含水多孔地层（储气）
aquifer[ˈækwɪfə]	n. 含水层，地下蓄水层
archives[ˈɑːkaɪvz]	n. 档案，档案室；案卷
as-built[əzˈbɪlt]	adj. 竣工；完工
atmospheric tank	常压罐
attached-to-bridge crossing	附桥跨越
attachment[əˈtætʃm(ə)nt]	n. 附件；附着物；附属装置
automatically[ˌɔːtəˈmætɪklɪ]	adv. 自动地；机械地；无意识地
available[əˈveɪləb(ə)l]	adj. 有效的，可得的；可利用的；空闲的
average absolute pressure	平均绝对压力

B

bare pipe laying	裸管敷设
be consistent with	与……一致；符合
be submitted to	被提交给……；服从于……
bending angle	弯曲角度
benzene[ˈbenziːn]	n. 苯
biomass[ˈbaɪə(u)mæs]	n. 生物量；生物质
black iron	平铁；黑铁板；黑钢板
bladder[ˈblædə]	n. 囊；泡
blasting engineering	爆破工程
block[blɒk]	n. 块；障碍物，阻碍；街区；
bottled supply station	瓶装供应站
branched pipeline network	枝状管网
branched[bˈrɑːntʃt]	adj. 分支的；分岔的；枝状的
brass[brɑːs]	n. 黄铜；黄铜制品
buffered user	缓冲用户
building service pipeline	用户引入管
buried trench laying	沟埋敷设

butane['bjuten]   n. 丁烷
butene['bjuːtiːn]   n. 丁烯

C

cargo['kɑːgəu]	n. 货物，船货
cast iron	铸铁
casualty['kæʒjuəltɪ]	n. 意外事故；伤亡人员
catering['keɪtərɪŋ]	n. 给养；提供饮食及服务
cathodic[ˌkæ'θodɪk]	adj. 阴极的；负极的
cathodic protection	阴极保护
cave storage	岩穴（储气）
CBG (coal bed gas)	煤层气
cellulose['seljuləuz;-s]	n. 纤维素；（植物的）细胞膜质
chloride['klɔraɪd]	n. 氯化物
circular['səːkjulə(r)]	adj. 圆形的；环形的
circular pipeline network	环状管网
clip sleeve	夹子套筒
close interval potential survey method (CIPS)	密间隔电位测试法
closure['kləuʒə(r)]	n. 闭合
collaboratively[kə'læbərətɪvlɪ]	adv. 合作地；协作
collecting well	集水井
combustible gas	可燃性气体
combustion[kəm'bʌstʃ(ə)n]	n. 燃烧，氧化；骚动
combustion potential	燃烧势
combustion-supporting gas	助燃气体
commodity[kə'mɔdɪtɪ]	n. 商品，货物；农产品；矿产品
convection[kən'vekʃən]	n. 对流，传送，传递，传导
completion and acceptance of the project	工程竣工验收
complex['kɔmpleks]	adj. 复杂的；复合的
compressed gas cylinders	压缩气筒
compression factor	压缩因子
compression[kəm'preʃən]	n. 压缩，浓缩；压榨，压迫
compressor[kəm'presə(r)]	n. 压缩机
compressor station	压气站
concealed work	隐蔽工程
concern[kən'sɜn]	n. 关系；关心；关心的事
	vt. 涉及，关系到；使担心

condensate drainage	凝水缸
condensate tank	冷凝槽
condensate[ˈkɒnd(ə)nseɪt]	n. 冷凝物；浓缩物
	adj. 浓缩的
	vt. 使浓缩；使压缩
condense[kənˈdens]	vi. 浓缩；凝结
confinement[kənˈfaɪnmənt]	n. 局限，限制；界限；约束
confliction[kənˈflɪkʃən;ˈkɒnflɪkʃən]	n. 冲突，抵触；矛盾，对立，分歧
confuse[kənˈfjuːz]	vt. 使混乱；使困惑
consortium[kənˈsɔːtɪəm]	n. 财团；联合；合伙
constant[ˈkɒnstənt]	n. 常数；常量
constitute[ˈkɒnstɪtjuːt]	vt. 组成，构成；建立；任命
consumption[kənˈsʌmpʃn]	n. 消耗；消耗量
continuity[ˌkɒntɪˈnjuːətɪ]	n. 连续；连续性
contradiction[ˌkɒntrəˈdɪkʃ(ə)n]	n. 矛盾；否认，反驳
controversial[ˌkantrəˈvɜːʃl]	adj. 有争议的；有争论的
conversion and production	置换投产
conversion ventilation	置换通气
convert[kənˈvɜːt]	vt. 使转变；转换…
corresponding[ˌkɒrəˈspɒndɪŋ]	adj. 相当的，对应的；符合的，一致的；相同的
	v. 相符合；类似；相配
corrode[kəˈrod]	vi. 受腐蚀；起腐蚀作用
	vt. 侵蚀；损害
corrosion resistance	耐腐蚀性能
coupling[ˈkʌplɪŋ]	n. 耦合；结合，联结
	v. 连接（couple 的 ing 形式）
critical[ˈkrɪtɪkl]	adj. 临界的
cryogenic[ˌkraɪəˈdʒenɪk]	adj. 冷冻的；低温学的；低温实验法的
current attenuation	电流衰减
cyclone[ˈsaɪkləun]	n. 旋风除尘器
cylinder[ˈsɪlɪndə]	n. 圆筒；气缸
cylindrical[səˈlɪndrɪkl]	adj. 圆柱形的；圆柱体的

D

debris[ˈdebriː]	n. 碎片，残骸；残渣
debugging[ˌdiːˈbʌgɪŋ]	n. 调试

decreasing[dɪˈkriːsɪŋ]	*adj.* 渐减的
	v. 减少；缩减使降低；降低
defect[ˈdɪfekt]	*n.* 缺点，缺陷；不足之处
deformation[ˈdɪfərˈmeʃən]	*n.* 变形
depleted oil and gas fields storage	枯竭的油气田（储气）
depleted[dɪˈplɪtɪd]	*adj.* 耗尽的；废弃的；贫化的
design flow	计算流量
design month	计算月
design pressure	计算压力
deterioration[dɪˌtɪrɪəˈreʃən]	*n.* 恶化；退化；堕落
diameter[daɪˈæmɪtə(r)]	*n.* 直径
diaphragm (or film) rupture	皮膜破裂
diaphragm[ˈdaɪəfræm]	*n.* 膈；隔膜，隔板
diffuser[dɪˈfjuːzə]	*n.* 扩压段，扩压器
dike[daɪk]	*n.* 堤
diolefin[dəɪˈəuləfɪn]	*n.* [化学] 二烯 (=diene)
direct acting regulator	直接作用式调压器
direct conversion	直接置换
direct current potential method	直流电流电位法
direct current voltage gradient (DCVG) technology	直流电压梯度技术
directional drilling laying	定向钻敷设
disadvantage[dɪsədˈvaːntɪdʒ]	*n.* 缺点；不利条件；损失
distribute[dɪˈstrɪbjuːt;ˈdɪstrɪbjuːt]	*vt.* 分配；散布；分开
distribution[ˌdɪstrɪˈbjuːʃn]	*n.* 分配；分布
distribution flow	途泄流量
domestic[dəuˈmestɪk]	*adj.* 家庭的，家用的
drainage[ˈdreɪnɪdʒ]	*n.* 排水
drilling leak detection	钻孔查漏
dry powder fire extinguisher	干粉灭火器
ductile cast iron pipe	球墨铸铁管道
ductile[ˈdʌktaɪl]	*adj.* 柔软的；易教导的；易延展的
dust collecting	除尘

E

eject[ɪˈdʒekt;ˈiːdʒekt]	*vt.* (从内部)排出；喷射，喷出
elastomer[ɪˈlæstəmə]	*n.* 弹性体，高弹体，（高）弹性塑料
elbow[ˈelbəu]	*n.* 弯头；弯管

emission[ɪˈmɪʃn]	n. 发射，散发；喷射；发行
emit[ɪˈmɪt]	vt. 发出，放射；发行；发表
equilibrium[ˌiːkwɪˈlɪbrɪəm;ˌekwɪ-]	n. 均衡，平静
equivalent[ɪˈkwɪvələnt]	n. 当量；等效
equivalent length	当量长度
ethane[ˈiːθeɪn;ˈeθ-]	n. 乙烷
evolution[ˌiːvəˈluːʃ(ə)n]	n. 演变；进化论；进展
excavation[ˌekskəˈveɪʃn]	n. 挖掘，发掘
excelsior[ekˈselsɪɔː]	adj. 精益求精的；不断向上的
exemption[ɪgˈzempʃn]	n. 免除，豁免；免税
exhaust[ɪgˈzɔːst]	vt. 排出；耗尽
	vi. 排气
expansion joint	补偿器
experimentally[ɪkˌsperəˈmentlɪ]	adv. 实验上；用实验方法
exploit[ɪkˈsplɔɪt]	n. 勋绩；功绩
	vt. 开发，开拓；剥削；开采
exploitation[ˌeksplɔɪˈteɪʃ(ə)n]	n.（资源等的）开发，开拓，开采；利用；剥削；广告推销
extensive[ɪkˈstensɪv]	adj. 广泛的；大量的；广阔的
external[ɪkˈstɜːnl]	adj. 外面的，外部的；外用的；外国的
external detection	外检测
extraction[ɪkˈstrækʃən]	n. 取出；抽出；拔出；抽出物

F

Fahrenheit[ˈfær(ə)nhaɪt]	n. 华氏温度计；华氏温标
	adj. 华氏温度计的；华氏的
feed stock	原料
ferocious[fəˈrəʊʃəs]	adj. 残忍的；惊人的
filament[ˈfɪləm(ə)nt]	n. 灯丝；细丝；细线；单纤维
filling station	灌装站
filter[ˈfɪltə(r)]	n. 过滤器
	vt. 过滤
filtration purification device	过滤净化装置
flammable[ˈflæməb(ə)l]	adj. 易燃的；可燃的；可燃性的
flange[flæn(d)ʒ]	n. 法兰；凸缘
flash hider	防火帽
flexibility[ˌfleksɪˈbɪlɪtɪ]	n. 灵活性；弹性；适应性

flexible['fleksəbl]	*adj.* 灵活的；易弯曲的；柔韧的
flow[fləu]	*n.* 流量；流动
fluctuation[ˌflʌktʃu'eɪʃn]	*n.* 波动，涨落，起伏
flue[fluː]	*n.* 烟道；暖气管
fracture['fræktʃə]	*n.* 破裂，断裂；骨折
	vi. 破裂；折断
frequency['friːkwənsɪ]	*n.* 频繁性；频率，次数
friction factor	摩擦阻力系数
frictional resistance	摩擦阻力
fundamentally[fʌndə'mentəlɪ]	*adv.* 根本地，从根本上；基础地
furnace['fəːnɪs]	*n.* 炉子，火炉；熔炉；蒸汽炉

G

gallon['gælən]	*n.* 加仑（容量单位）
galvanized['gælvənaɪzd]	*adj.* 镀锌的，电镀的
galvanized steel pipe	镀锌钢管
gas appliance	燃具
gas consumption	用气量
gas fueling station	加气站
gas meter	燃气计量表
gas metering device	用气计量装置
gas mixing station	混气站
gas pipeline and tube bundle	管道和管束（储气）
gas quality testing equipment	气质检测设备
gas sendout	供气量
gas storage tank	储气罐（储气）
gas underground reservoir	地下储气库（储气）
gasholder['gæshəuldə]	*n.* 煤气库，气柜，贮气罐
gate station	门站
generous['dʒen(ə)rəs]	*adj.* 慷慨的，大方的；宽宏大量的；有雅量的
gradient['greɪdɪənt]	*n.* 梯度；坡度；倾斜度
	adj. 倾斜的；步行的
gravity['grævɪtɪ]	*n.* 重力，地心引力
grind[graɪnd]	*vt.* 磨碎；磨快

H

handwheel['hændwi:l]	n. 手轮，操纵轮
hazardous['hæzədəs]	adj. 有危险的；冒险的
heat insulation case	保温情况
hedgerow['hedʒrəu]	n. 绿篱
helium['hi:liəm]	n. 氦（符号为 He）
high-density['haɪ'densɪtɪ]	adj. 高密度的
horizontal['hɔrɪ'zɔntəl]	adj. 水平的，地平的
horizontal cylindrical tank	卧式圆筒形罐
horizontal main pipeline	水平干管
horsepower['hɔ:spauə]	n. 马力（功率单位），功率
hulls[hʌlz]	n. 外壳
hydraulic fracturing	水力压裂
hydraulic[haɪ'drɔ:lɪk]	adj. 液压的；水力的；水力学的
hydrocarbon[,haɪdrə(u)'ka:b(ə)n]	n. [有化]碳氢化合物；烃
hydrogen['haɪdrədʒən]	n. [化学]氢
hypothesis[haɪ'pɒθəsɪs]	n. 假设；假说；假定

I

immune[ɪ'mjun]	adj. 免疫的；免于……的，免除的 n. 免疫者；免除者
impact[ɪm'pækt]	n. 影响；效果；碰撞；冲击力
impeller[ɪm'pelə]	n. 叶轮，叶轮片，转子
implement['ɪmplɪm(ə)nt]	n. 工具，器械；家具；手段 vt. 实施，执行；使生效，实现；把……填满
imprecisely[,ɪmprɪ'saɪzlɪ]	adv. 不严密地
impurities[ɪm'pjuərɪtɪs]	n. 不纯；不洁；杂质
in full swing	活跃；蓬勃高涨
in the routing inspection	在巡检过程中
in the vent	在排气口
indirect acting regulator	间接作用式调压器
indirect conversion	间接置换
indoor gas pipeline	室内燃气管道
indoor open installation	明管敷设

industrial[ɪnˈdʌstrɪəl]	adj. 工业的，产业的；供工业用的
inert gas	惰性气体
inevitably[ɪnˈevɪtəblɪ]	adv. 不可避免地；必然地
inflation[ɪnˈfleʃən]	n. 膨胀；通货膨胀；夸张；自命不凡
infrastructure[ˈɪnfrəstrʌktʃə(r)]	n. 基础设施；基础建设
insufficient[ɪnsəˈfɪʃ(ə)nt]	adj. 不足的，不充足的
insulating flanges	绝缘法兰
insulation performance	绝缘性能
integrity[ɪnˈtegrɪtɪ]	n. 完整，正直；诚实；廉正
interface[ˈɪntəˌfes]	n. 界面；接口；交界面
internal detection	内检测
invisible[ɪnˈvɪzəbl]	adj. 无形的，看不见的；无形的
isolation[ˌaɪsəˈleɪʃən]	n. 隔离；分离；脱离；孤立
isothermal[ˌaɪsəuˈθɜːməl]	adj. 等温的；恒温的
iteration[ˌɪtəˈreɪʃn]	n. 迭代

J

joint venture	合资企业；联合经营

K

kinematic[ˌkɪnɪˈmætɪk]	adj. 运动的；运动学的

L

laminar[ˈlæmɪnə]	adj. 层流的
	n. 层流
landfill[ˈlændˌfɪl]	n. 垃圾填埋地；垃圾堆
leak plugging	堵漏
leak test	严密性试验
legislation[ˌledʒɪsˈleɪʃən]	n. 立法；法律
liable[ˈlaɪəbl]	adj. 有义务的；应受罚的；有……倾向的；易……的
liaison letter	联络单
linear loss	沿程损失
lineup[ˈlaɪnˌʌp]	n. 阵容；一组人
liquefied hydrocarbon	液化烃

liquefied natural gas storage	液化天然气储存
liquid-state storage	液态储气
local resistance	局部阻力
loop analysis method	回路分析法
looped network	环状管网
looped[luːpt]	adj. 环状的；成圈的
low pressure piston-type tank	低压干式罐
low pressure water-sealed tank	低压湿式罐
low temperature and atmospheric pressure storage	低温常压储存
lubrication[luːbrɪˈkeɪʃən]	n. 润滑；润滑作用

M

make great contributions to	对…做出贡献
malfunction[mælˈfʌŋ(k)ʃ(ə)n]	n. 故障；失灵；疾病
malignant[məˈlɪgnənt]	adj. 恶性的；有害的；有恶意的
mandatory[ˈmændətəri]	adj. 强制的；命令的；受委托的
manufactured[mænjuˈfæktʃəd]	adj. 人造的；人工的
material of pipe	管材
matrix[ˈmeɪtrɪks]	n. 矩阵
maximum uneven factor of daily consumption	日高峰系数
maximum uneven factor of monthly consumption	月高峰系数
meander type regulator	曲流式调压器
mercaptan[məˈkæpt(ə)n]	n. 硫醇
metering device	计量装置
methane[ˈmiːθeɪn; ˈmeθeɪn]	n. 甲烷；沼气
metropolitan[ˌmetrəˈpɒlɪtən]	adj. 大都市的；大主教辖区的
modification[ˌmɒdɪfɪˈkeɪʃn]	n. 修正；修改
modified[ˈmɒdɪfaɪd]	adj. 改进的，修改的；改良的
moisture[ˈmɒɪstʃə]	n. 水分；湿度；潮湿；降雨量
molecular[məˈlekjulə]	adj. 分子的；由分子组成的
monitoring and control system	监测与控制系统
monoxide[məˈnɒksaɪd]	n. 一氧化物
monthly average daily gas consumption	月平均日用气量
moratorium[ˌmɒrəˈtɔːrɪəm]	n. 暂停，中止；延期偿付
mould[məuld]	vi. 发霉
multi-stage network system	多级管网系统

N

naphthalene[ˈnæfθəliːn]	n. 卫生球；臭樟脑；萘
natural gas storage and distribution station	天然气储配站
NGV(Natural Gas Vehicle)	天然气车辆
nitrogen[ˈnaɪtrədʒən]	n. 氮
node[nəud]	n.（计算机网络的）节点
nominal diameter	公称直径
normative[ˈnɔːmətɪv]	adj. 规范的，标准的
nozzle[ˈnɒz(ə)l]	n. 喷嘴；管口；鼻

O

odorant[ˈəud(ə)r(ə)nt]	n. 添味剂，臭味剂；有气味的东西
	adj. 有气味的；有香气的
odorization[ˌəudəraɪˈzeɪʃən]	n. 加臭
odorization device	加臭装置
oil resources	石油资源
oil-indexed	与石油挂钩的
olefin[ˈəulɪfɪn]	n. 烯烃
orientate[ˈɔːrɪənteɪt]	vi. 向东，定向
	vt. 给……定位；使适应
originally[əˈrɪdʒənəlɪ]	adv. 最初，起初；本来
output[ˈautput]	n. 输出，输出量；产量；出产
	vt. 输出
oven[ˈʌv(ə)n]	n. 炉，灶；烤炉，烤箱
overhead crossing	架空跨越
overhead gas pipeline	架空燃气管道
overhead laying	架空敷设

P

parameter[pəˈræmɪtə(r)]	n. 参数
peak shaving	调峰
peaking[ˈpiːkɪŋ]	n. 剧烈增加；脉冲修尖；求峰值
pharmaceutical[ˌfɑːməˈsuːtɪk(ə)l;-ˈsjuː-]	adj. 制药（学）的
perforation[ˌpɜːfəˈreʃən]	n. 穿孔；贯穿

peripheral[pəˈrɪfərəl]	adj. 周围的；外围的
perish[ˈperɪʃ]	vi. 死亡；毁灭；腐烂
	vt. 使麻木；毁坏
perpendicular[ˌpəːp(ə)nˈdɪkjulə]	adj. 垂直的，正交的；直立的
	n. 垂线；垂直的位置
pesticide[ˈpestɪsaɪd]	n. 杀虫剂
pharmaceutical[ˌfaːməˈsuːtɪk(ə)l;-ˈsjuː-]	n. 药物
pier[pɪə(r)]	n. 码头，防波堤；桥墩
pigging[ˈpɪgɪŋ]	n. 清管；清管作业
pigging device	清管装置
pillar[ˈpɪlə]	n. 柱子，柱形物，栋梁；墩；支柱
pilot[ˈpaɪlət]	n. 飞行员；领航员；指挥器
pipe attachment	管道附件
pipe jacking laying	顶管敷设
pipeline bridge crossing	管桥跨越
piping concealment	暗封敷设
piping embedment	暗埋敷设
piston type regulator	活塞式调压器
piston[ˈpɪst(ə)n]	n. 活塞
pneumatic[njuːˈmætɪk]	adj. 气动的；充气的
polyethylene[ˌpɒlɪˈeθəlɪn]	n. 聚乙烯
polyethylene pipeline	聚乙烯管道
polyvinyl[ˌpɒlɪˈvaɪn(ə)l]	adj. 乙烯聚合物的
polyvinyl chloride	聚氯乙烯
	n. 聚乙烯化合物
poppet[ˈpɒpɪt]	n. 提升阀，盘形活门
potential deviation	电位偏移
predetermined pressure	预定压力
predominantly[prɪˈdɑmɪnəntlɪ]	adv. 主要地；显著地
preliminary[prɪˈlɪmɪnərɪ]	adj. 初步的
premise[ˈpremɪs]	n. 前提
prescribe[prɪˈskraɪb]	vi. 建立规定，法律或指示；开处方
	vt. 指定，规定；指定，规定
pressure head	压头
pressure level	压力级制
pressure rating	压力等级
pressure regulating device	调压装置
pressure tank	压力罐

pressure test	试压
pressurized['preʃəraɪzd]	adj. 加压的；受压的；增压的；加压的
	v. 增压；密封；使加压
pressurized equipment	加压设备
profit['prɒfɪt]	n. 好处，益处；（财产等的）收益
propagation[ˌprɒpəˈgeɪʃən]	n. 传播；繁殖；增殖
propane['prəupeɪn]	n. 丙烷
pump[pʌmp]	n. 泵，抽水机；打气筒
purification[ˌpjuərɪfɪˈkeɪʃn]	n. 净化；洗净；提纯
purification device	净化装置
put in place	落实到位

R

raffinate['ræfɪneɪt]	n. 残油液；剩余液
rail tanker transportation	铁路槽车运输
range[reɪndʒ]	n. 范围；幅度；排；山脉
rated pressure/normal operating pressure	燃气额定压力
rationality[ˌræʃəˈnælətɪ]	n. 合理性；（复数）合理的行动
reciprocating[rɪˈsɪprəˌkeɪtɪŋ]	adj. 往复的；交互的
redundant[rɪˈdʌndənt]	adj. 过多的，过剩的；多余的
refers to	指的是
refinery[rɪˈfaɪnərɪ]	n. 精炼厂；提炼厂；冶炼厂
refinery wastewater	炼油污水；炼油废水
refuel[riːˈfjuəl]	vt. 补给燃料
regularity[ˌregjuˈlærətɪ]	n. 规律；规律性
regulator['regjuleɪtə]	n. 调节器，稳流器，调节阀
regulator cabinet (box)	调压柜（箱）
regulator station	调压站
relatively['relətɪvlɪ]	adv. 相当地；相对地，比较地
relief[rɪˈliːf]	n. 减轻；解除；缓解；泄放；安慰；浮雕；救济
resistance[rɪˈzɪstəns]	n. 阻力
respectively[rɪˈspektɪvlɪ]	adv. 各自地；各个地；分别地
restrictive[rɪˈstrɪktɪv]	adj. 限制的；限制性的；约束的
rigorous['rɪg(ə)rəs]	adj. 严格的，严厉的；严密的；严酷的
riser['raɪzər]	n. 立管；起义者；叛乱者
road tanker transportation	公路槽车运输

rotary[ˈrəutərɪ]	adj. 旋转的；转动的
roughness[rʌfnəs]	n. 粗糙
routine[ruːˈtiːn]	adj. 例行的；常规的；日常的；普通的
routing inspection	巡线；巡检
rubber gasket	胶垫
rush repair	抢修

S

safety first, precaution crucial	安全第一，预防为主
sagging pipe	下垂管
salt layer storage	盐矿层（储气）
scatter[ˈskætə(r)]	vi. 撒开；分散
	vt. （使）散开，（使）分散
seal test	密封试验
seasonal uneven gas consumption	季节用气不均衡性
self-operated regulator	自力式调压器
service pipe	引入管
sewer[ˈsuːə;ˈsjuː]	n. 下水道；阴沟
shift system	轮班制度
significance[sɪɡˈnɪfɪk(ə)ns]	n. 意义；重要性；意思
simultaneous[sɪm(ə)lˈteɪnɪəs]	adj. 同时的；同时发生的；联立的
slender[ˈslendə]	adj. 细长的；苗条的；微薄的
sludge[ˈslʌdʒ]	n. 烂泥；泥泞；泥状雪；沉淀物
slurry[ˈsləːrɪ]	n. 悬浮液，泥浆；水泥浆
soil collapse	土壤塌陷
solid crystalline hydrate	固体结晶水化物
solid-state storage	固态储存
spherical[ˈsfɪərɪkəl]	adj. 球形的，球面的；天体的
spherical tank	球形罐
spring[sprɪŋ]	n. 春天；弹簧；泉水；活力；跳跃
starch[stɑːtʃ]	n. 淀粉；刻板，生硬
state-backed	国家支持的
statutory[stætjut(ə)rɪ]	adj. 法定的；法令的
statutory exemption	法定豁免
steady[ˈstedɪ]	adj. 稳定的；稳态的
steel skeleton polyethylene composite pipe	钢骨架聚乙烯复合管道
stokehole pipeline	炉前管道

strength test	强度试验
structure['strʌktʃə]	n. 结构；构造；建筑物
surge chamber	调压室
synonymous[sɪ'nɒnɪməs]	adj. 同义词的；同义的

T

T type regulator	T 型调压器
t-branch	三通
tar[tɑː]	n. 焦油；柏油；水手
technical disclosure	技术交底
tedious['tiːdɪəs]	adj. 繁琐的；冗长的
tensile['tensl]	adj. 拉力的；可伸长的；可拉长的
terminal['tɜːmɪnl]	adj. 末端的
	n. 终端；末端
termination[tɜːmɪ'neɪʃ(ə)n]	n. 结束，终止
tetrahydrothiophene[ˌtetrəˌhaɪdrə'θaɪəfiːn]	n. 四氢噻吩 (THT)
thermodynamic[ˌθəːməʊdaɪ'næmɪk]	adj. 热力学的；使用热动力的
thermostat['θɜːməstæt]	n. 恒温器；自动调温器
thioethers[ˌθaɪəʊ'iːθə]	n. 硫醚
thiol['θaɪəʊl]	n. 硫醇
thiophene['θaɪəfiːn]	n. 噻吩
throttle valve	节流阀
total hub	总枢纽
total valve	总阀门
tracer leak detection method	示踪剂检漏法
transit flow	转输流量
transmission[trænsˈmɪʃən]	n. 传输；传送；播送；传递；传导；传达
transoceanic[ˌtrænsəʊʃɪ'ænɪk]	adj. 横越海洋的；在海洋彼岸的
tremendous[trə'mendəs]	adj. 极大的，巨大的；惊人的
turbulent['tɜːbjələnt]	adj. 紊流的；湍流的
	n. 紊流；湍流
turf[tɜːf]	n. 草坪；草皮

U

underground laying	地下敷设
underground storage	地下储气

undertaken[ˌʌndəˈteɪkən]	v. 从事；开始进行
underwater crossing	水下穿越
uneven[ʌnˈiːv(ə)n]	adj. 不均匀的；不平均的
uneven factor of daily consumption	日不均匀系数
uneven factor of hourly consumption	小时不均匀系数
user branched pipeline	用户支管
utilities[juːˈtɪlɪtɪz]	n. 公共事业；实用工具

V

valve chamber maintenance	阀室维护
vaporizing station	气化站
variable[ˈveərɪəbl]	adj. 变化的；可变的
vehicle[ˈviːəkl]	n. 车辆；工具；交通工具
velocity[vəˈlɒsətɪ]	n. 速率，速度
vent pipe	放散管
venting[ˈventɪŋ]	n. 排气；通气 v. 排放（vent 的现在分词）
ventilation[ventɪˈleɪʃən]	n. 通风；换气，空气流通
vertical[ˈvɜːtɪk(ə)l]	adj. 垂直的，直立的
vessel[ˈves(ə)l]	n. 船，舰；脉管，血管；容器，器皿
vibration[vaɪˈbreʃən]	n. 振动；犹豫；心灵感应
viscosity[vɪˈskɒsɪtɪ]	n. 黏性；黏度
visual inspection	外观检查
volumetric[vɔljuˈmetrɪk]	adj. 容量的，容积的；体积的

W

water jacket	[机械学] 水套；冷却管
waterway transportation	水路槽船运输
weir[wɪə]	n. 堰；低坝
weld joint	焊缝
Wobbe index	沃伯指数；华白数

Translations and Answers

任务 1 城市燃气分类

一、听力部分

城镇燃气指符合燃气质量规范，供给民用、商用和工业生产的燃料用公用性质燃气，一般包括天然气、液化石油气、人工燃气和生物质气（沼气）。

天然气是在油田、天然气田和煤层中发现的一种气态可燃矿物。它是数百万年前动植物死亡后的有机体逐渐转化形成的。天然气的主要成分是甲烷，还包含乙烷、丙烷和丁烷等烃类气体，并包含其他非烃类气体。

液化石油气是在开采天然气、石油或石油炼制过程中作为副产品而获得的燃气，它是城镇燃气的主要气源之一。液化石油气的主要成分是丙烷、丙烯、丁烷和丁烯。

人工燃气是指以固体或液体可燃物为原料，经各种热加工生产出的可燃气体。一般根据制气原料和加工方式不同，可将人工燃气分为煤制气和油制气两类。

各种有机物质，如蛋白质、纤维素、脂肪、淀粉等，在隔绝空气的条件下发酵，并在微生物的作用下产生的可燃气体，称为沼气（生物气），可分为天然沼气和人工沼气两大类。

1)fuel 2)natural 3)coal 4)component 5)gases
6)by-product 7)combustible 8)oil 9)fat 10)artificial

二、口语部分

有用的句子

（1）按照气源的起源和生产方式的不同，可以将燃气分为天然气、人工燃气、液化石油气和生物质气四类。

（2）液化石油气既是我国民用、商用、工业生产、机动车用户的重要燃料，也是化工生产中的重要原料。

（3）生物质气俗称"沼气"，可分为天然沼气和人工沼气两大类。

（4）天然沼气是自然界中有机质自然形成的沼气，如污泥沼气，阴沟沼气，矿井、煤层产出的沼气等。

（5）人工沼气是一种可再生能源。

（6）燃气热值是城镇燃气质量分析中的重要指标。

（7）无论是天然气、液化石油气还是人工燃气，由于产地不同，即使是同一种类的燃气，其成分和热值也不尽相同，有时区别还可能很大。

（8）燃气按热值高低习惯上可分为高热值燃气、中等热值燃气和低热值燃气。

（9）在燃气性质中影响燃烧特性的参数主要有沃泊指数（又译华白数）和燃烧势。

（10）沃泊指数是一项反映燃具热负荷恒定状况的指标，可以分析、控制燃气的互换性。

（11）燃烧势是一项反映燃具燃烧稳定状况的综合指标，能更全面地判断燃气的燃烧特性。

三、阅读部分

城市燃气的基本性质

燃气是由多种可燃和不可燃的单一气体组成的混合气体，其主要性质包括物理性质、热力学性质和燃烧特性。

燃气的主要物理性质：燃气的主要物理性质通常指燃气的平均相对分子质量、密度和相对密度、黏度、露点、沸点、临界参数等。

燃气的热力学性质：燃气的热力学性质包括汽化潜热、比热容、导热系数等。汽化潜热是指单位质量（1kg）的液体变成其处于平衡状态的蒸气所吸收的热量。在不发生相变和化学反应的条件下，单位质量物质温度升高 1K 所吸收的热量，称为该物质的比热容（或比热）。导热系数是物质导热能力的特性参数，表示沿着导热方向每米长度上的温度降低 1K 时每小时所传导的热量。气体的导热系数随温度和压力的升高而增大。

燃气主要有以下 5 个方面的燃烧特性。

（1）热值：单位体积（$1m^3$）或单位质量（1kg）的燃气完全燃烧所能放出的全部热量称为燃气的发热量。燃气热值是评价燃气质量的重要指标之一，也是正确选用燃烧设备或燃具时所必须考虑的一项质量指标，它将燃气分为高热值和低热值两种类型。

（2）理论空气需要量与烟气量：众所周知，燃气燃烧需要供给适量的氧气，氧气过多或过少都对燃烧不利。燃烧所需要的氧气一般都是从空气中直接获得的。理论空气需要量是指每标准立方米燃气按燃烧反应方程式计算完全燃烧时所需的空气量。燃气的热值越高，燃烧所需理论空气量也越多。燃烧同样体积的液化石油气、天然气和焦炉煤气所需的空气量是不同的。液化石油气所需的空气量约为燃烧天然气所需空气量的 3 倍，为焦炉煤气所需空气量的 6 倍。燃气燃烧后的产物就是烟气。当供给理论空气量时，燃气完全燃烧后产生的烟气量称为理论烟气量。理论烟气的组分是二氧化碳、二氧化硫、氮气和水蒸气，不完全燃烧时的烟气中还会有一氧化碳。

（3）着火温度：任何可燃气体在一定条件下与氧接触都要发生氧化反应，由稳定的氧化反应转变为不稳定的氧化反应而引起燃烧的一瞬间，即称为着火。着火温度是指可燃气体与空气（或氧气）的混合物开始进行燃烧反应的最低温度。

（4）爆炸极限：可燃气体与空气（或氧气）的混合物能发生着火以致引起爆炸的浓度范围，称为爆炸极限，其最低浓度称为爆炸下限，最高浓度称为爆炸上限。

（5）火焰传播速度：垂直于燃烧焰面，火焰向未燃气体方向传播的速度称为火焰传播速度，也称为燃烧速度或燃烧率。它与燃气性质、燃气—空气混合物的组成、温度和压力有关，是判定燃气互换性的基本参数。

（6）沃泊指数和燃烧势：沃泊指数是在互换性问题产生初期所使用的一个互换性判定指数。当置换气和被置换气（基准气）的化学性质、物理性质和燃烧特性变化较大时，燃气的燃烧速度会发生较大的变化。此时只用沃泊指数已不能控制燃气的互换性，需采用燃烧势来控制燃气的互换性。燃烧势反映了燃气燃烧火焰所产生的离焰、黄焰、回火和不完全燃烧的倾向性，是一项反应燃具燃气燃烧稳定状况的综合指标。

四、翻译部分

城市燃气发展概况

城镇燃气作为城镇基础设施的重要组成部分，不仅关系到人们生活质量的提高、城镇自然环境和社会环境的改善，而且已日益成为国民经济中具有先导性、全局性的基础产业。

与发达国家相比，我国城镇燃气起步较晚。概括来讲，我国现代城镇燃气产业的发展大致经历了以下三个阶段。

第一阶段：20世纪80年代以前，在国家钢铁工业大发展的带动下和国家节能资金的支持下，全国建成了一批城镇燃气利用工程，许多城镇建设了燃气管网设施。在这一阶段，以发展煤制气为主，取得了普及用户、增加燃气供应量的成绩。

第二阶段：20世纪80年代至90年代前期，液化石油气（Liquefied Petroleum Gas，简称LPG）和天然气得到了很快的发展，形成了多种气源并存的格局。同时出现了国内现有资源难以满足城镇发展和经济建设需求的状况。广东等沿海经济较发达但能源缺乏的地区开始进口液化石油气。至此，国内外液化石油气资源得到了较充分的利用，液化石油气成为我国城镇燃气的主要气源之一。

第三阶段：20世纪90年代后期，随着我国能源结构的挑战，以陕甘宁天然气进京为代表的天然气供应拉开了序幕，特别是涩—宁—兰、西气东输和忠—武输气管道等国家重点天然气工程的相继建成投产，为城镇管道燃气的大发展提供了前提，并提供了前所未有的发展机遇。这也标志着我国城镇燃气的天然气时代已经到来。压缩天然气（compressed natural gas，简称CNG）、国产液化天然气（liquefied natural gas，简称LNG）和进口液化天然气的共同增进，更是为我国城镇燃气多气源的发展和利用带来了新的前景。

任务 2 城市燃气需求

一、听力部分

随着燃气事业的发展，特别是天然气的大量开采与利用，燃气已成为能源供应的重要支柱。根据用户用气特点，城镇燃气用户一般包括以下几种类型。

（1）居民用户：居民用户是指以燃气为燃料进行炊事和制备热水的家庭燃气用户。它是城镇燃气供气的基本用户之一，必须保证连续稳定的供气。

（2）商业用户：商业用户是城镇燃气供气的又一个基本用户，它是指用于商业设施或公共建筑制备热水或炊事的燃气用户，包括职工食堂、饮食业、幼儿园、托儿所、医院、宾馆酒店、理发店、浴池、洗衣店、超市、机关、科研单位、大学、中学、小学等机构的用气。在学校和科研单位，燃气除了用于炊事、热水、淋浴外，还用于实验室。

（3）工业企业生产用户：工业企业生产用户是以燃气为燃料从事工业生产的用户。这类用户用气量较大，比较有规律。

（4）采暖、通风和空调用户：采暖、通风和空调用户是指以燃气为燃料进行采暖、制冷的用户，属于季节性负荷，因此，给这类用户供气时，还必须具备有效调节季节不均匀用气的措施。

（5）燃气汽车用户：燃气汽车用户是以燃气作为汽车动力燃料的燃气用户。发展燃气汽车是降低城镇大气污染的有效措施之一。燃气汽车与燃油汽车相比，燃料价格具有较明显的优势。

（6）其他用户：其他用户主要包括两部分，一部分是由于外力破坏、自然腐蚀、用户使用不当、生产放空等因素造成的管网漏损量；另一部分是因发展过程中出现没有预见到的新情况而超出了原计算的供气量。

（7）热电站用气，当电站采用城镇燃气为燃料调峰发电时，城镇燃气负荷还应包括电站用气量。将直接使用低污染燃烧的天然气转换为零污染物排放的电能来使用，这是天然气应用的一大发展方向。

有资料显示，天然气已开始在鲜花和蔬菜的暖棚种植、粮食烘干与储藏、农副产品的深加工、生物、医药、农药、燃气燃料电池灯等方面得到应用和开发，这必将导致城镇燃气用户规模的不断扩大和分类的进一步细化。

1)energy 2)household 3)stable 4)public 5)regular 6)seasonal
7)power 8)leakage 9)load 10)planting

二、口语部分

有用的句子

(1) 我国城镇燃气原来主要用于居民生活和一般工商业，现在正向燃气发电、燃气空调、燃气汽车、以化工为主的工业等应用领域迅速扩展。

(2) 单个居民用户的用气量不大，用气随机性较强。

(3) 商业用户用气量不是很大，用气比较有规律。

(4) 工业企业生产用户用气量较大，比较有规律。

(5) 采暖通风和空调用户具有突出的不均匀用气的特点，但他们在采暖期内的用气是相对稳定的。

(6) 燃气汽车用户用气量与城镇燃气汽车的数量及运营情况有关，受季节等外界因素影响较小。

(7) 热电站用气量非常大，这部分气量不包括在规划城镇燃气管网的用气量当中，需按独立项目考虑。

三、阅读部分

城市燃气的质量要求

城市燃气在进入输配管网和供给用户前，都应满足热值相对稳定、毒性小和杂质少等基本要求，并且要达到一定的质量指标并保持质量的相对稳定，这对于保障城市燃气输配系统和用户用气的安全，减少管道腐蚀和堵塞，降低对环境的污染，保障系统的经济合理性等都具有重要的意义。

城市燃气的主要杂质：由于城市燃气气源的来源和制取方法的不同，其中所含杂质也不尽相同。例如人工燃气中的主要杂质有焦油、粉尘、苯、萘、氨、硫化氢和氧化氮等；天然气中的主要杂质有硫化氢、水、凝析油、粉尘等；液化石油气中的主要杂质有硫化氢、水、二烯烃、残液等。

城市燃气的质量指标：为了保证燃气用具在其允许的适应范围内工作，提高燃气的标准化水平，要求城市燃气质量指标应符合下列要求：

(1) 城市燃气的发热量和组分的波动应符合城市燃气互换的要求；

(2) 城市燃气偏离基准气的波动范围宜按现行的《城市燃气分类和基本特性》（GB/T 13611—2006）规定，并应适当留有余地。

城市燃气的加臭：城市燃气在空气中应具有嗅觉能力一般的正常人可以察觉的臭味，以警觉燃气的泄露，及时采取措施，消除隐患。当城镇燃气自身气味不能使人有效察觉和明显区别于日常环境中的其他气味时，应进行补充加臭。我国目前采用的臭味剂主要是四氢噻吩（THT），又称噻吩烷，为无色、无毒、无腐蚀性的透明油状液体，具有恶臭气味。四氢噻吩是用于燃气加臭最为稳定的化合物，与其他加臭剂如硫醇类、硫醚类相比，具有抗氧化性能强、化学性质稳定、不污染环境等优点。

四、翻译部分

城市燃气的生产

从气井井口采出或从矿场分离器分离出的天然气含有不同数量的、在大气条件下处于液相的较重烃类，并含有水蒸气、硫化物（如硫化氢）、二氧化碳、氮和氦等非烃类气体，一般不适宜大多数用户直接使用。它们大多需要经处理以脱除不希望有的组分（如硫化氢、水蒸气）后方可作为商品天然气供城镇用户使用。

液化天然气是由天然气液化制取的以甲烷为主的液烃化合物。一般是在常压下将天然气冷冻到约 -162℃ 而变为液体，以有利于输送和储存。

压缩天然气是经过压缩的高压商品天然气，其主要成分是甲烷。由于它不仅抗爆性能和燃烧性能好，燃烧产物中的温室气体及其他有害物质含量也很少，而且生产成本较低，因此是一种很有发展前途的汽车优质替代燃料。

液化石油气是在开采天然气、石油或石油炼制过程中作为副产品而获得的燃气，是我国城镇燃气的主要气源之一。按其来源可分为炼厂液化石油气和油气田液化石油气两种。前者是由炼油厂的二次加工过程所得，主要由丙烷、丙烯、丁烷和丁烯组成。后者则是由天然气加工过程所得到的，主要成分为丙烷或/和丁烷，但不含烯烃。

任务 3　城市燃气管网分类

一、听力部分

燃气管道的分类

首先，城市燃气管道根据输气压力分为高压 A 和高压 B 燃气管道、次高压 A 和次高压 B 燃气管道、中压 A 和中压 B 燃气管道、低压燃气管道。其次，城市燃气管道根据用途分为输气管道、配气管道、用户引入管道、室内燃气管道和工业企业燃气管道。第三，城市燃气管道根据敷设方式分为地下燃气管道和架空燃气管道。第四，城市燃气管道根据管网形状分为环状管网、枝状管网和环枝状管网。

1)according to 2)sub-high 3)divided 4)transmission 5)indoor 6)industrial 7)overhead 8)laying 9)shape 10)circular

二、口语部分

1．有用的句子

（1）在燃气输配系统中，各种压力级制的燃气管道之间应通过调压装置相连。

（2）居民用户和小型商业用户一般直接由低压管道供气。

（3）高压燃气通过调压站才能送入中压管道和流程中需要高压燃气的大型工厂企业。

（4）高压 A 输气管道通常是贯穿省、地区或连接城镇的长输管线，它有时也构成城市输配管网系统的外环网。

（5）城镇燃气管网系统中各级压力的干管应连成环网。

（6）输气管道主要用于输送城镇燃气，一般是高压、次高压燃气干管，或是连接大型工业企业的专供管道。

（7）配气管道包括街区的和庭院的分配管道，一般为中压、低压燃气管道。

（8）室内燃气管道是指通过用户管道引入管的总阀门将燃气引向用户室内并分配到每个燃气用具的管道。

2．对话

A：王教授，早上好！很高兴见到您！

B：早上好！很高兴见到您！

A：听说您最近参加了关于燃气管网信息化管理技术的讲座。

B：印象非常深刻啊！通过这次讲座，我学到了很多。

A：太好了！我最近正在写一篇这方面的论文，可以问您一些问题吗？

B：好的，没问题！

A：目前，在城市燃气行业有哪几种信息化管理技术？

B：有 SCADA 系统、GIS 和 MIS 等。

A：您能解释一下吗？

B：好的。SCADA 系统是数据采集与监视控制系统。GIS 是地理信息系统，MIS 是管理信息系统。

A：在燃气行业信息管理系统的应用越来越重要了。

B：是啊，尤其是 SCADA 系统。它可以应用于城市燃气输配系统、电力系统、给水系统、石油化工系统等。

A：非常感谢，今天我收获很大。

B：不用客气。

三、阅读部分

城市燃气管网系统

　　城镇燃气管网系统根据所采用的管网压力级制不同,可分为一级管网系统、两级管网系统、三级管网系统和多级管网系统。一级管网系统是指仅由一种压力等级管道组成的管网系统,一般有低压一级管网、中压 A 一级管网和中压 B 一级管网三种配置。两级管网系统是指具有两种压力等级管道组成的管网系统,多由低压两级管道和中压两级管道组成。三级管网系统是指包含三种压力等级的管道组成的输配系统,一般由低压三级管道、中压三级管道和高(或次高)压三级管道组成。含有三种以上压力等级的管网系统称为多级管网系统。

　　在选择城市燃气管网系统时,应考虑以下主要因素:

　　(1) 气源情况,包括燃气的种类和性质、供气规模、供气压力等。

　　(2) 城镇现状与发展规划,包括城镇街区和道路、建筑特点、人口密度、用户情况(类型、数量、分布、用气压力、用气量、气化率、供气原则)及其发展规划等。

　　(3) 大型用户与特殊用户状况,包括数目、分布、用气压力、用气量、生产工艺特点等。

　　(4) 储气措施与储气能力。

　　(5) 城镇地理环境以及天然和人工障碍物情况,包括土壤状况(质地、腐蚀性、气温、冰冻线等),桥梁、河流、湖泊、铁路等的数量、分布情况,地下管线和地下建构筑物的现状及其改建、扩建规划。

　　(6) 所需管材、管道附件、调压设备等的生产、供应情况。

四、翻译部分

　　(1) In the city gas pipeline wiring, we must consider the basic situation of urban development planning and the gas pressure in the pipeline, etc.

　　(2) Outdoor gas pipeline that mostly adopts underground laying should be laid along urban roads, sidewalks, or in the greening area.

　　(3) Indoor gas pipeline in general adopts **indoor open installation**, specially piping embedment and piping concealment.

　　(4) In the city gas pipeline wiring, according to the specification requirements we determine the position of the pipeline plane and vertical section, and design through and across the obstacles.

　　(5) The region which the pressure is more than 1.6MPa gas pipeline through, by the intensity along the buildings, should be divided into four levels, and according to the level, we make the corresponding pipeline design and layout requirements.

(6) 压力不大于1.6MPa的燃气管道涵盖了压力分级中的次高压、中压和低压三个级别，其主要功能是向用户输气和通过调压站向低压环网配气。

(7) 燃气管道穿越铁路和电车轨道时，必须采用保护套管或混凝土套管，并要垂直穿越。

(8) 燃气管道在水下穿越河流的敷设方法有沟埋敷设、裸管敷设、顶管敷设和定向钻敷设。

(9) 燃气管道的跨越一般有附桥跨越、管桥跨越、架空跨越等。

(10) 居民用户的室内燃气管道可按压力大小分为低压进户和中压进户两类。

一、听力部分

城市燃气的储运

一般而言，天然气常用的储存方式有储气罐储气、地下储气库储气、管道和管束储气、液化天然气储存。储气罐储气分为低压储气罐储气和高压储气罐储气。天然气的地下储存通常包括枯竭的油气田储气、含水多孔地层储气、盐矿层储气和岩穴储气等。液化天然气储存目前一般采用低温常压储存的方法。天然气的其他储存方式有在低温液化石油气中的储存和固态储存等。城市燃气的运输方式有管道运输、铁路槽车运输、公路槽车运输和水路槽船运输。

1) commonly 2) underground 3) liquefied 4) depleted 5) porous 6) At present
7) atmospheric 8) petroleum 9) modes 10) waterway

二、口语部分

1. 有用的句子
(1) 按其原理和特点，天然气的储存方式大致可分为低压储气和高压储气等。
(2) 按储存形式，天然气的储存方式可分为气态储存和液态储存等。
(3) 枯竭的油气田是最好的和最可靠的地下储气库。

（4）天然气的固态储存是将天然气在一定的压力和温度下转变成固体的结晶液化物，储存于钢制的储罐中。

（5）按地质构造划分，地下储气库分为枯竭油气田型、含水层型、盐穴型、岩洞型及废弃矿井型。

（6）低压储气罐有干式罐和湿式罐两大类。

（7）高压储气罐是一种容积罐，常用的两种型式为卧式圆筒形罐和球形罐。

（8）高压燃气管束储气和长输管道末端储气是平衡小时不均匀用气的有效办法。

（9）与地上储气设施相比，地下储气库具有容量大、适应性强、经济性好、安全度高、占地面积少、环境影响小等一系列优点。

（10）车间燃气管道宜明设，架设高度应不低于两米。

2．对话

A：李教授，早上好！我正在自学天然气储存知识，但是我遇到了一些困难。我可以问您一些问题吗？

B：当然可以了！

A：谢谢您！我在书中看到了一个不熟悉的术语。那就是 ANG。

B：嗯，ANG 是天然气的吸附储存，它是一种天然气的储存方法。

A：我明白了。您能给我介绍一下天然气吸附储存（ANG）方面的一些知识吗？

B：好的。天然气吸附储存(ANG)是在储罐中装入固体吸附剂(如沸石、分子筛、硅胶、炭黑、活性炭等)，以使其在一定储存压力(3.5～6.0MPa)下吸附天然气，达到与压缩天然气（CNG）相接近的存储容量。

A：天然气的吸附储存有哪些优点呢？

B：天然气吸附储存的优点在于储存压力相对较低，对储气和充气设备耐压性能要求低，投资费用低，而且安全性能可靠，日常操作维护方便，运营费用低。

A：现在我对 ANG 有了一些了解，真是太感谢您了！

B：不用客气。

三、阅读部分

工业企业燃气供应系统

工业企业燃气输配系统通常由工厂引入管、厂区燃气管道、车间燃气管道、工厂总调压站或车间调压装置、用气计量装置、安全控制装置和炉前管道等构成。工业企业用户一般由城镇中压或次高压管网供气，用气量小且用气压力为低压的用户可直接由低压管网供气，其最佳方案的选择应由技术经济分析结果决定。大型工业企业可敷设专用管道，与城镇燃气门站或长输管线连接。

工业企业燃气管网系统可概括地分为一级管网系统和两级管网系统两类。选择工业企业燃气管网系统时应考虑以下主要因素：

（1）连接引入管处的城镇燃气分配管网的燃气压力。

(2) 各用气车间燃烧器前所需的燃气额定压力。
(3) 用气车间在厂区分布的位置。
(4) 车间的用气量及用气规模。
(5) 与其他管道的关系，管理检修条件和经济效果。

厂区燃气管道布线时应遵循下列原则：
(1) 厂区管道一般采用钢管。
(2) 燃气引入管宜设在使用燃气的房间或燃气表间内。
(3) 厂区燃气管道可以采用地下敷设，也可以采用架空敷设。
(4) 厂区燃气管道的末端应设放散管。
(5) 厂区架空燃气管道系统应尽可能简单而明显，以便于施工安装、操作管理和日常维修。

车间燃气管网系统有枝状和环状两种，一般采用枝状系统，环状系统只用于特别重要的车间。

四、翻译部分

(1) Indoor gas pipeline system generally consists of building service pipeline, horizontal main pipeline, riser, user branched pipeline, gas meter, appliance connecting pipeline and gas appliance.

(2) According to the setting mode of gas meter, indoor gas pipeline can be divided into two categories — scattered meter and concentrated meter.

(3) Building service pipeline refers to the pipeline between the outdoor distribution branched pipeline and the total valve of user indoor gas inlet.

(4) Gas riser can't be laid in the bedroom or bathroom.

(5) Residential gas branched pipeline can adopt piping embedment or piping concealment setting when it can't be indoor open installation.

(6) 燃气支管穿过墙壁时应安装在套管内。
(7) 高层建筑物自重大，沉降量显著，易在引入管处造成破坏。
(8) 高层建筑燃气立管的管道长、自重大，需在立管底部设置支墩。
(9) 用具连接管（又称下垂管）是在燃气支管上连接燃气用具的垂直管段。
(10) 室内燃气管道的管材应采用低压流体输送钢管，并应尽量采用镀锌钢管。

任务 5 燃气管网调峰

一、听力部分

城市燃气的供求平衡

城市的用气量随时间而变化，每月、每日和每时都不相同，而气源供气量一般变化不大，尤其是使用长距离输气管道时。因此，供气量和用气量经常发生不平衡。为了保证按用户要求供气，必须解决供气量与用气量的不平衡问题。解决月（季度）间、日间或小时间不均衡用气问题，可采用不同的储气方法，如地下储气、储气罐储气、液态储气和输气管道末段储气等。地下储气主要用于克服季节用气不均衡性，储气罐储气、液态储气和输气管道末段储气主要用于解决小时用气不平衡问题。

1)varies 2)especially 3)Therefore 4)requirements 5)unbalance 6)uneven
7)such as 8)liquid-state 9)overcome 10)solve

二、口语部分

1．有用的句子

（1）为了解决均匀供气和不均匀耗气之间的矛盾，必须采取合适的方法使燃气输配系统供需平衡。

（2）各种储气设施的主要作用就是在用气量小于供气量时，将多余气体储存起来，以弥补用气量大于供气量时的不足。

（3）用气高峰时，会出现供不应求的现象；用气低峰时，会出现供过于求的现象。

（4）城镇燃气属于整个天然气系统的下游，长输管道为中游，天然气开采、净化为上游。

（5）采用低温液态储存，适合于远距离输运，通常储存量都很大，否则经济性差。

（6）储气罐是指专门用来储存气体的设备。

（7）利用长输干管储气或城镇外环高压管道储气是最经济的一种方法，也是国内外最常用的一种方法。

(8) 城镇燃气供应的缓冲用户是指能使一年中的用气波动达到最小值的可中断用气的用户和非高峰期用气用户。

2．对话

A：苏教授，下午好！

B：下午好！很高兴见到您！

A：我有一些问题要向您请教，恐怕要占用您一些时间。您能给我介绍一下城市燃气供需不平衡的原因吗？

B：没问题。城市的用气量随时间而变化，而气源供气量一般变化不大。因此，供气量和用气量之间经常发生不平衡。

A：为了保证按用户的需求供气，我们必须解决供气量与用气量的不平衡问题。

B：是的。目前，人们尽力采取各种方法解决供气量与用气量的不平衡问题。

A：您能简要地向我解释一下调整城市燃气供需平衡的方法吗？

B：嗯，第一种方法是改变气源的生产能力和设置机动气源。这种方法必须考虑气源运转、停止的难易程度，气源生产负荷变化的可能性和变化幅度。同时，我们应考虑供气的安全可靠性和技术经济合理性。

A：我明白了。除了这种方法，还有其他的方法吗？

B：是的，还有两种方法。利用缓冲用户进行调节供气量与用气量的不平衡问题。这种方法主要是利用一些大型的工业企业和锅炉房等作为缓冲用户，我们可用此方法平衡季节不均匀用气及部分日不均匀用气。另一种方法是利用储气设施进行调节供气量与用气量的不平衡问题。

A：非常感谢，今天我收获很大。

B：不用客气。

三、阅读部分

储气容积的计算

1．储气罐容积的确定

对城镇燃气输配系统中所需储气罐容量的计算，按气源及输气能否按日需用气量供气，区分为两种工况：

(1) 供气能按日需用气量变化时，储气罐容量应按计算月的计算日 24h 的燃气供需平衡条件进行计算。

(2) 供气不能按日需用气量变化时，储气罐容量应按计算月平均每周 168h 的燃气供需平衡条件进行计算。

确定储气罐容积的一般具体步骤如下：

(1) 确定月高峰系数。

(2) 确定日不均匀系数和日高峰系数。

(3) 确定时不均匀系数。

(4) 计算居民用户和商业用户的年平均日用气量。

(5) 计算居民用户和商业用户的月平均日用气量。

(6) 计算居民用户和商业用户一周内某天的小时用气量。

(7) 计算工业用户的小时用气量。

(8) 计算用气量累计值。

(9) 计算供气量累计值。

(10) 计算小时储气量。

(11) 用列表法对 24h 或一周的每小时两种条件下分别进行上面相关量的计算。

(12) 所需的储气容积等于最大小时储气量与最小小时储气量的绝对值之和。

2. 长输管线末端储气量计算

长输管线末端储气是利用燃气长距离输送系统的最末一个压气站与城市门站之间的管段进行储气。其储气能力为储气终了时与储气开始时该段输气管线中存气量之差，可按下列步骤近似计算。

(1) 根据终点用户要求的最低供气压力和正常输气量确定储气开始时起点压力 $p_{1\min}$，有

$$p_{1\min} = \sqrt{p_{2\min}^2 + \frac{9.053 Q_V^2 \lambda Z_{\min} STL}{d^5 \times 10^7}}$$

式中 $p_{1\min}$——输气管道在储气开始时起点的绝对压力，MPa；

$p_{2\min}$——输气管道在储气开始时终点的绝对压力，MPa；

Q_V——气体（$p_0 = 0.101325 \text{MPa}, T_0 = 293\text{k}$）的流量，m³/d；

λ——水力摩阻系数；

$Z_{\min}$——气体在储气开始时管道内平均压缩因子；

S——气体的相对密度；

T——输气管道内气体的平均温度，K；

L——输气管道的长度，km；

d——输气管道内径，cm。

(2) 根据压气站的最高工作压力或管线强度允许压力以及正常输气量确定储气终了时的管线终点压力 $p_{2\max}$，有

$$p_{2\max} = \sqrt{p_{1\max}^2 - \frac{9.053 Q_V^2 \lambda Z_{\max} STL}{d^5 \times 10^7}}$$

式中 $p_{2\max}$——输气管道在储气结束时终点的绝对压力，MPa；

$p_{1\max}$——输气管道在储气结束时起点的绝对压力，MPa；

$Z_{\max}$——气体在储气结束时管道内平均压缩因子。

(3) 确定储气开始时的平均绝对压力 $p_{m.\min}$，有

$$p_{m.\min} = \frac{2}{3}\left(p_{1\min} + \frac{p_{2\min}^2}{p_{1\min} + p_{2\min}}\right)$$

式中 $p_{m.\min}$——输气管道在储气开始时管道内平均绝对压力，MPa。

(4) 确定储气终了时的平均绝对压力 $p_{m.max}$，有

$$p_{m.max} = \frac{2}{3}\left(p_{1max} + \frac{p_{2max}^2}{p_{1max} + p_{2max}}\right)$$

式中　$p_{m.max}$——输气管道在储气结束时管道内平均绝对压力，MPa。

(5) 确定输气管道的储气能力 V，有

$$V = \frac{V_c T_0}{p_0 T}\left(\frac{p_{m.max}}{Z_{max}} - \frac{p_{m.min}}{Z_{min}}\right)$$

式中　V——输气管道（$p_0 = 0.101325\text{MPa}, T_0 = 293\text{k}$）的储气能力，$m^3$；
　　　V_c——输气管道的几何容积，m^3。

四、翻译部分

(1) In order to ensure gas uninterrupted supply according to the requirements of users, we must consider the balance problem of the gas supply and use.

(2) When adjusting the supply and demand balance of city gas, usually the upstream gas supply side works out the seasonal gas supply and demand balance, and the downstream gas town works out the daily gas supply and demand balance.

(3) When gas city is not too far away from the place of origin of natural gas, we can adopt to adjust the supply of gas well to balance part of the daily uneven gas consumption.

(4) According to the gas consumption and gas conditions of industrial enterprises, residents living and business, we formulate the scheduling plan to adjust gas sendout.

(5) Gas underground reservoir has large capacity, low cost and operating costs, and can be used to balance the seasonal and part of the daily uneven gas consumption.

(6) 在长输管线供气到来之后，供需双方应明确彼此在调峰和安全供气方面所承担的责任。

(7) 在整个城市燃气输配系统中，需要从全局来解决调峰问题，以求得天然气系统的优化，达到经济合理的目的。

(8) 液化天然气气化方便，负荷调节范围广，适于调节各种不均匀用气。

(9) 金属储气罐对于保证燃气管网的正常运行是必不可少的。

(10) 随着天然气日益普及，在大、中城市天然气门站普遍采用高压储气球罐，用作调峰城市气源。

任务 6 燃气管网水力计算

一、听力部分

计算机在水力计算中的应用

城市燃气管网水力计算是城市燃气设计的主要工作之一。管网水力计算常用的方法是回路分析法和节点流量法（又称为水力计算法）。回路分析法只适用于小型的枝状管网，节点流量法具有在不知道管段流量的情况下通过迭代逼近真解的特点，适用于各种大型复杂管网，但该法计算工作量大，用手工计算非常困难，通常在计算机上进行。它以AutoCAD平台为基础，采用VB二次开发技术，实现燃气管网管段节点坐标的读取，自动生成燃气管网水力计算图，并可采用VC++可视化编程语言编制城市燃气管网水力计算软件。

1) loop 2) branched 3) iteration 4) flow 5) coordinates 6) hydraulic 7) generated

二、口语部分

1. 有用的句子

（1）随着计算机技术的迅速发展和普及，克服了以往手工计算燃气管网时工作量大的困难。

（2）水力计算是城市燃气设计的主要工作之一。

（3）通过水力计算，来确定管道的投资和金属消耗量，并保证管网工作的可靠性。

（4）在实际工程中，产生局部阻力处的流动常处于紊流的粗糙区。

（5）管段的总阻力包括摩擦阻力和局部阻力。

（6）管网基本上可分为枝状管网和环状管网。

（7）室内燃气管道是指从引入管到管道末端燃具前的管道。

（8）沿管段直接分配给用户的流量称为途泄流量。

（9）任何管网图形都是由一些节点和管段连接起来的几何图形。

（10）表达管网图形性质的有力工具是矩阵，可以用关联矩阵和节点间关系矩阵来表示管网图形。

2．对话

A：城市燃气管网水力计算是城市燃气设计的主要工作之一。

B：没错！设计时要求燃气管网既满足使用的需要，又省投资。

A：除此之外，对运行中的燃气管网，应保证合理的生产调度、可进行管网事故模拟，并可建立处置预案，可对管网事故进行紧急处理。

B：我们仅说了设计原则，那么，它的具体任务是什么呢？

A：简单地说，燃气管网水力计算的任务是根据燃气的计算流量和允许的压力降来确定管径。

B：公式的推导有哪些条件呢？

A：在通常情况下的一小段时间内，燃气管道中的燃气流动可视为稳定流。

B：还有其他的么？

A：假设燃气在管道中是等温流动，则λ（燃气管道摩擦系数）和T（设计采用的燃气温度）均为常数；考虑管道压力变化不太大，Z（压缩因子）也可视为常数。

B：哦。让我想一想。在实际计算中，直接用公式应该比较繁琐吧？

A：你说的没错！通常为了简化计算，将推导出的水力计算公式绘制成水力计算图。

B：需要进行参数修正吗？

A：是的！由于计算图是在燃气特定的参数下绘制的，当实际参数与计算图上的参数不同时，要进行修正。

B：好的。非常感谢你的解答！使我学到了更多关于燃气管网水力计算的知识。

A：不用客气。

三、阅读部分

燃气管道的水力计算

正确地进行水力计算，是关系到输配系统经济性和可靠性的问题，是城市燃气规划与设计中的重要环节。

1．城市燃气管道水力计算公式和计算图表

1）低压燃气管道水力计算公式

考虑燃气在不同材质管道、不同流动状态条件下，低压燃气管道单位长度的摩阻损失具体采用下列各式计算：

$$\frac{\Delta p}{L}=1.13\times10^{10}\frac{Q}{d^4}\nu\rho\frac{T}{T_0} \quad \text{（层流区，}Re<2100\text{）}$$

$$\frac{\Delta p}{L}=1.88\times10^6\left(1+\frac{11.8Q-7\times10^4 dv}{23Q-1\times10^5 dv}\right)\frac{Q^2}{d^5}\rho\frac{T}{T_0} \quad \text{（临界区，}Re=2100\sim3500\text{）}$$

$$\frac{\Delta p}{L} = 6.89 \times 10^6 \left(\frac{K}{d} + 192.2\frac{dv}{Q}\right)^{0.25} \frac{Q^2}{d^5} \rho \frac{T}{T_0} \qquad (紊流区，Re>3500，钢管、塑料管)$$

$$\frac{\Delta p}{L} = 6.39 \times 10^6 \left(\frac{1}{d} + 5158\frac{dv}{Q}\right)^{0.284} \frac{Q^2}{d^5} \rho \frac{T}{T_0} \qquad (紊流区，Re>3500，铸铁管)$$

式中　Re——雷诺数；
　　　Δp——燃气管道摩阻损失，Pa；
　　　L——燃气管道计算长度，m；
　　　Q——燃气管道计算流量，m³/h；
　　　d——管道内径，mm；
　　　ρ——燃气密度，kg/m³；
　　　T——设计中所采用的燃气温度，K；
　　　T_0——取值为273.15K；
　　　v——燃气在0℃和101.325kPa条件下的运动黏度，m²/s；
　　　K——管壁内表面的当量绝对粗糙度，mm。

2）高、中压燃气管道水力计算公式

当对燃气管道的摩阻系数采用手工计算时，根据燃气管道的不同材质，高、中压燃气管道单位长度摩阻损失宜采用下列各式计算：

$$\frac{p_1^2 - p_2^2}{L} = 1.4 \times 10^9 \left(\frac{K}{d} + 192.2\frac{dv}{Q}\right)^{0.25} \frac{Q^2}{d^5} \rho \frac{T}{T_0} \qquad (钢管、塑料管)$$

$$\frac{p_1^2 - p_2^2}{L} = 1.3 \times 10^9 \left(\frac{1}{d} + 5158\frac{dv}{Q}\right)^{0.284} \frac{Q^2}{d^5} \rho \frac{T}{T_0} \qquad (铸铁管)$$

式中　p_1——燃气管道起点压力（绝对压力），kPa；
　　　p_2——燃气管道终点压力（绝对压力），kPa；
　　　L——燃气管道计算长度，km。

3）燃气管道水力计算图表

压力不同、管材不同，水力计算公式也不同，所以也就对应着不同的水力计算图表。另外，由于不同种类燃气的密度、黏度等有很大的不同，所以计算图表也不同。

决定水力计算图表的因素主要有三个，即不同的燃气种类、管道的压力级别、不同的管道材质。三者的不同组合得到不同的水力计算图表。

4）附加压头

由于空气与燃气密度不同，当管道始、末端存在标高差时，在燃气管道中将产生附加压头。对始末端高程差值变化甚大的个别管段，包括低压分配管道及建筑物的室内的低压燃气管道，必须将附加压头计算在内。

附加压头的计算公式为

$$\Delta p = g\left(\rho_a - \rho_g\right)\Delta H$$

式中　Δp——附加压头，Pa；
　　　ρ_a——空气密度，kg/m³，取 ρ_a =1.293；
　　　ρ_g——燃气密度，kg/m³；
　　　g——重力加速度，m/s²，取 g=9.81；
　　　ΔH——管道末端与始端的标高差值。

$\rho_a > \rho_g$ 时，即管道内流动气体上升时将产生一种升力，下降时将增加阻力。

$\rho_a < \rho_g$ 时，即管道内流动气体下降时将产生一种升力，上升时将增加阻力。

5）局部压力损失计算

当燃气流经三通管、弯管、变径异型管、阀门等管路附件时，由于几何边界的急剧改变，燃气在管道内的气流方向和气流断面改变，燃气运动受到扰乱，必然产生额外的压力损失。

一般而言，对于城市燃气管网，管网的局部损失一般以沿程损失的 5%～10% 估计。

对于室内燃气管道和厂、站区域的燃气管道，由于管路附件较多，局部损失所占的比例较大，应进行计算。计算方法有两种，一种是用公式计算，另一种用当量长度法。

公式法计算对应的公式为

$$\Delta p = \Sigma \xi \frac{v^2}{2} \rho$$

式中　Δp——管段局部阻力损失，Pa；
　　　$\Sigma \xi$——计算管段中局部阻力系数的总和；
　　　v——燃气流动速度，m/s。

当量长度法计算对应的公式为

$$\Delta p = \Sigma \xi \frac{v^2}{2} \rho = \lambda \frac{L_d}{d} \frac{v^2}{2} \rho$$

$$L_d = \Sigma \xi \frac{d}{\lambda}$$

式中　λ——水力摩阻系数；
　　　L_d——管段局部摩阻的当量长度，m；
　　　d——管道内径，m。

2. 枝状管网

枝状管网的水力计算步骤为：

(1) 对管网的节点和管道进行编号。

(2) 根据管线图和用气情况，确定管网各管段的计算流量。

(3) 选定枝状管网的干管，根据给定的允许压力降确定管线单位长度上的允许压力降。

(4) 根据管段的计算流量及单位长度允许压力降预选管径。

(5) 根据所选定的标准管径，反算管段实际沿程压力降和局部压力降，并计算总的压力降。

（6）检查计算结果。若总的压力降未超过允许值并趋近允许值，则认为计算合格，否则应适当变动管径，直到总压力降小于并尽量趋近允许值为止。

3．环状管网

1) 流动规律

（1）节点连续性方程；

（2）环压降闭合差为零。

2) 水力计算步骤

环状管网水力计算通常采用解管段方程组、解环方程组和解节点方程组的方法。它是对压降方程、连续性方程和能量方程的联立求解。

（1）绘制管网平面图，对节点、管段、环网进行编号，并标明管道长度、集中负荷、气源或调压站的位置。

（2）计算管网各管段的途泄流量。

（3）按气流沿最短路径从供气点流向零点的原则，拟定环网各管段中的燃气流向。气流的方向总是流离供气点，而不应逆向流动。

（4）从零点开始，逐一推算管段的转输流量。

（5）求管网各管段的计算流量。

（6）预选管径：

①由管网的允许压降及供气点至零点的计算长度确定单位长度允许的压降；

②由单位长度允许压力降预选管径；

③将管径化为规格系列。

（7）初步计算管网各管段的压降及每环的压力降闭合差；

（8）计算管网平差，求每环的校正流量，使所有封闭环网压力降的代数和等于零或接近于零，达到工程容许的误差范围；

（9）验证总压降，若满足要求，计算结束；不满足，重复第（6）、（7）、（8）步。

四、翻译部分

(1) When doing the hydraulic computation of indoor gas pipeline, we should designate and arrange user gas appliances and draw up piping diagram, firstly according to planar graph and profile map of buildings.

(2) The coincidence factor method is used to calculate and determine the design flow of every pipe section.

(3) The maximum value of the pressure drop from service pipe to every gas appliance is the systematic pressure drop.

(4) To write the serial number of pipe section according to the order, any change location of pipe diameter and flow should be numbered, and the actual length of every calculated section should be labelled.

(5) The pipe diameter of every pipe section is set according to the design flow.

(6) 根据燃气的种类、密度和运动黏度选择水力计算图,并确定管段单位长度的压降值。

(7) 计算各管段的附加压力。

(8) 计算各管段的实际压力损失。

(9) 求室内燃气管道的总压力降时,人工燃气计算压力降一般不超过 80 ~ 100Pa。

(10) 将室内燃气管道的总压力降与允许的压力降进行比较时,如不合适,则可调整个别管段的管径。

一、听力部分

天然气储配站工艺流程

储配站采取两级调压,高压储气,次高压供气工艺。天然气由气矿配气站输入工厂储配站的高压干管时的压力为 0.5 ~ 1.5MPa,经储配站调压后输入工业和民用的次高压供气管网时的压力为 0.3MPa。当储配站在低峰负荷时,进站天然气经计量调压,直接向次高压管网供气,同时向储气罐充气。当进站气压低于储气罐压力时,需经压缩机增压至储气罐压力后,再向储气罐充气,一般储气罐压力为 0.8 ~ 1.5MPa。当用户处于高峰负荷时,储气罐应放出天然气,经调压并进入供气管道。

1)distribution 2)civil 3)directly 4)aeration 5)compressor 6)tank 7)pipeline

二、口语部分

1. 有用的句子

(1) 储配站是将储气站和配气站合并。

(2) 储配站按压力可分为高压储配站和低压储配站。

(3) 储气站的主要设施是储气罐。

(4) 门站是长输管线终点配气站。

(5) 输气管道试压前应采用清管器进行清管,并应不少于两次。

(6) 采用不停气密闭清管流程。

(7) 除尘设备有两种，分别是旋风除尘器和过滤器。

2．对话

A：张明，早上好！很高兴见到你！

B：早上好，小丽！很高兴见到你！听说你去参加了关于燃气输配的讲座。

A：嗯，印象非常深刻啊！通过这次讲座，我学到了很多。

B：太好了！最近，我正在写关于燃气输配方面的论文，能请教你一些问题么？

A：好的，没问题。

B：在燃气储配站，有一项工艺是加臭，请问它的主要作用是什么啊？

A：为了安全使用天然气，当天然气从管道和设备中泄漏出来时，加臭能使人马上察觉到，同时也是一种管道检漏的方法。

B：目前，国内外城镇燃气常用的加臭剂是四氢噻吩。

A：没错。一般每1000方天然气中要加入16～32g四氢噻吩。

B：好的，我明白了！非常感谢你！今天真是学到了非常重要的知识！

A：不用客气！明天见！

B：再见！

三、阅读部分

门站工艺流程

门站负责接收气源（包括煤制气厂、天然气、煤层气及有余气可供应用的工厂等，供城镇使用的燃气），进行计量，检测质量，按城镇供气的输配要求，控制并调节向城镇供应的燃气流量与压力，必要时还要对燃气进行净化、加臭。

门站站址的选择：

(1) 门站与周围建筑的防火间距必须符合现行国家标准《建筑设计防火规范》的规定；

(2) 门站站址应具有适宜的地形、工程地质、供电、给排水和通信等条件；

(3) 门站站址宜靠近城镇用气负荷中心地区。

门站的工艺流程是天然气由高压管道输送到门站总进口阀，进入汇管，再经三路（两开一备）过滤计量；计量后，通过两路并联调压系统（一开一备），将压力由6.0MPa调到4.0MPa后，经加臭装置进入地下输气管道，进入加气站。

四、翻译部分

(1) Classification of tank area:

① Aboveground tank area of flammable liquid (atmospheric tank area);

② Aboveground tank area of liquefied hydrocarbon, flammable gas, combustion-supporting gas (pressure tank area).

(2) Atmospheric tank is the one whose design pressure is no more than 6.9kPa (gauge pres-

sure of tank top).

(3) Processing units, a group of storage tanks and loading area which may send out flammable gas or factory sewage treatment plant should be arranged on the windward side of annual minimum frequency direction in the personnel centralized place and with the open fire or sending out spark.

(4) A group of storage tanks of liquefied hydrocarbon or flammable liquid should not be arranged close to flood drainage trench tightly.

(5) Low–pressure tank of natural gas is the one whose design pressure is more than 6.9kPa and less than 0.1MPa (gauge pressure of tank top).

(6) 压力罐是设计压力大于或等于0.1MPa（罐顶表压）的储罐。

(7) 罐区泡沫站应布置在罐组防火堤外的非防爆区，与可燃液体罐的防火间距不宜小于20m。

(8) 工艺装置或可燃气体、液化烃、可燃液体的罐组与周围的消防车道之间不宜种植绿篱或茂密的灌木丛。

(9) 在可燃液体罐组防火堤内可种植生长高度不超过15cm、含水分多的四季常青的草皮。

(10) 液化烃罐组防火堤内严禁绿化。

燃气场站设备与设施

一、听力部分

储气设施

城市的天然气需求量并不是不变的，它随着时间而变化。但是气源的供应几乎是不变的，它并不随时间而变化。为了解决不变的供应和多变的需求之间的矛盾，城市的天然气配气系统通常都会设有储气站。储气站在低需求时储存气体，在高需求时释放出气体。下面是几种常见的储存方式。在需求量低时，气体被泵入到地下有适宜地质结构的孔隙内。在需求量高时，气体从地下孔隙中释放出来。这种适宜的地质结构通常是枯竭的油田或者气田、地下含水多孔结构、盐矿层和洞穴等。

最经济的方法就是把城市天然气储存在枯竭的油田或气田内。地下的储集层可以储存大量的天然气，同时它的投资和运行费用均很低。这种办法节约了上千吨的钢材。它通常用来满足季节性调峰或是部分日调峰。对于城市燃气来说它是一种理想的储存方法。困难的是怎样在城市的附近找到既经济又适宜的地质结构储存天然气。我国第一批地下库容量为10,000,000 立方米。自从西气东输开始后，大规模的地下储气库开始修建。液化天然气的体积要小得多，其体积约为同质量的天然气体积的 1/600。它可以储存在一个绝热的容器内。在天然气需求量高峰时，液化天然气将被气化，以供给用户使用。

由于液化天然气在国家间很容易进行运输和贸易，现在它已经成为一种国际性的商品。最近，世界上液化天然气的生产和贸易活跃起来。液化天然气在石油和天然气行业蓬勃发展。

高压管道和长输管道的末段储气是小时调峰的有效方式。高压管道组被埋藏在地下，在其内部的天然气被压缩成高压气体进行储存。在夜间天然气需求量低的时候长输管道末段可以储存一定数量的天然气，当白天天然气需求量高的时候释放天然气。

在中国最常用的天然气储存方法是用低压或高压储气罐来储存气体，以平衡日或小时的天然气需求量的波动。与其他储存方法相比，储气罐储存会消耗更多的金属材料，而且花费更高。但是当其他储气方法没法使用时，储气罐储气是用来储存气体的唯一方式。

1)storage　2)peak　3)pumped　4)steel　5)heat　6)active　7)pressure

二、口语部分

有用的句子

（1）天然气通常以液体状态储存在 −162℃的环境中，其目的就是减少体积、方便运输。

（2）压缩天然气通常以 24.821MPa 的压力储存在圆柱形的钢罐中。

（3）燃气的含碳量比煤和石油的含碳量低。因此，燃气燃烧时释放出的 CO_2 比煤和石油少。

（4）燃气易于点燃，而且在高燃烧率下，其燃烧易于控制。

（5）天然气是煤和液化石油气中含碳量最低的，所以被称为绿色燃料。

（6）城市天然气供应的发展是中国 21 世纪能源战略的一个重要部分。

（7）城市天然气分配系统负责将天然气安全的从气源输送到居民用户、商业用户和工业用户。

（8）城市天然气分配系统是把天然气供应给城市成千上万用户的关键部分。

（9）气体调压器的主要功能是匹配通过调压器的气流和系统设置的气体需求量。

（10）调压器必须保持系统压力在某一可接受的范围内。

（11）大多数调压器有三种基本的操作元件：一个加载装置、一个传感元件和一个控制元件。

（12）按照定义，压缩机是用来压缩气体物质的。

（13）一旦确定了吸入压力、排出压力、吸入气体温度、需要的流量和气体的组成，就能选择合适的压缩机。

（14）离心式压缩机是一种动力机器。它能够产生从整体轴叶轮中获得能量的连续流动的流体。

（15）天然气进入到压缩机站，然后通过干燥剂去压缩机。

（16）液化天然气站的花费均不相同，它取决于每小时处理的燃料量和该地储存的数量。

三、阅读部分

压力调节器

压力调节器在家庭和工业中应用广泛。例如，在气体烤架上使用压力调节器可以调节丙烷压力，在家庭壁炉中使用可以调节天然气压力，在医学/牙医设备中可调节氧气和麻醉气体，在气体自动化系统中可以调节压缩空气，在引擎中可以调节燃料压力，在燃料电池中可以调节氢气压力。虽然这些应用的变化很大，但压力调节器起到的作用是一样的。压力调节器将供给压力调到一个更低的出口压力，并且无论入口压力如何波动都保持这个出口压力不变。这种压力上的降低是压力调节器的重要特征。出口压力总是低于入口压力。

当选择压力调节器时，需要考虑很多因素。重要的考虑因素包括材料的选取、操作压力（入口和出口压力）、流量要求、所使用的流体（气体、液体、危险的或是惰性的）、温度等等。

材料选取：可供选择的多种不同的材料包括但不仅限于抗腐蚀不锈钢、黄铜、铝和塑料。不锈钢的寿命长，并且适用于干净的房间和腐蚀性流体。当花费是主要的影响因素时，黄铜、铝和塑料的调压器是最佳选择。铝的重量轻，同时塑料适用于涉及体液的多种医疗设备。当需要一次性物品时，塑料产品通常是理想的。

操作压力：在选择最佳的调压器之前，入口和出口压力是重要的考虑因素。需要回答的重要问题包括入口压力的波动范围是多少？需要的出口压力是多少？出口压力的允许偏差是多少？

流量需求：设备需要的最大流量是多少？流量的变化范围是多少？出口的要求也是重点考虑因素。

所使用的流体（气体、液体、危险的或是惰性的）：在决定你的设备的最佳材料之前，考虑流体的化学性质也是重要的。每一种流体都有其独特的特性，所以在选择与工艺流体相接触的合适的材料时要谨慎小心。实际上确定流体是否可燃或有危险也是十分重要的。将非放散调节器用于有危险的、易爆的或昂贵的气体是比较好的，因为这时没有气体被允许排放到大气中。放散调节器将会释放出过量的下游压力到空气中。根据所有的安全规程，过量的流体应被安全排放。

温度：被选用为压力调节器的材料不仅需要与流体相匹配，还要能够在预期的工作温度下正常运行。首先要考虑的是所选择的这个弹性体在全部预期的温度范围内是否能正常运行。另外，在极限的应用情况下工作温度可能影响设备的流量和/或弹簧的刚度。

运行中的压力调节器：压力调节器包括三个主要元件，即降压或限压元件（通常是一个提升阀）、传感元件（通常是薄膜或活塞）和负载元件（通常是一个弹簧）。在运行时，弹簧产生了打开阀门的力。压力从进气口引入后，流过阀门，并且压在了传感装置（薄膜或活塞）上。调整后的压力作用在传感元件上，产生了与弹簧力相反的力，然后关闭阀门。

四、翻译部分

压缩机

因为天然气输送管道从气田开始延伸了很长的距离，所以压缩机是必不可少的。当天然气没有足够的流动势能时，压缩机站是必需的。在天然气生产工业中通常有五种类型的压缩机站被利用：

（1）现场天然气集气站从压力不足以产生流入输送或分配系统所需的流量的井中收集气体。这些集气站通常处理的入口压力从低于大气压力到 5.17MPa，并且每天转输体积为几千到数百万立方英尺。

（2）中继站或干线站来提高输送管线的压力。他们通常在 1.38MPa 到 8.96MPa 的压力范围内压缩大体积的气体。

（3）再加压或循环站为工艺或二次采油项目提供 41.37MPa 的气体压力。

（4）现场储气站用来压缩干线气体，以注入储气井中的压力达到 27.58MPa。

（5）分输站是用来把气体从气源以大约 0.14MPa 到 0.69MPa 泵送至中压或高压分输管线，或者泵送到瓶中以 17.24MPa 的压力进行储存。

压缩机的类型：目前，天然气生产工业中所使用的压缩机分为两种不同的类型：往复式压缩机和旋转式压缩机。

往复式压缩机

在天然气工业中，往复式压缩机是最常见的。被建造的往复式压缩机适合于所有的压力和容积。与旋转式压缩机相比，往复式压缩机有更多的运转部件，因此机械效率较低。往复式压缩机的每一个缸塞系统都包括有一个活塞、气缸、气缸盖、吸入阀和排出阀，还包括将旋转运动转换为往复式运动的其他必需部件。往复式压缩机是通过选择气缸内的合适的活塞位移和余隙容积，共同确定一定范围内的压缩比而设计的。这种余隙容积可以是固定的或可变的，这取决于操作范围和需要的载荷变化量的程度。典型的往复式压缩机可以在排出压力达到 68.95MPa 时，传送高达 30,000 立方英尺/分钟（cfm）的气体体积流量。

旋转式压缩机

旋转式压缩机分为两类：离心式压缩机和旋转鼓风机。

离心式压缩机包括具有流道的外壳、装有叶轮的旋转轴、轴承和防止气体沿着轴逸出的密封装置。离心式压缩机的运转部件不多，因为仅仅只有叶轮和轴旋转。因此，它的效率高，并且其润滑油的消耗和维护的费用都低。由于其低压缩率和低摩擦损耗，所以通常不需要使用冷却水。离心式压缩机的压缩率较低，这是因为没有正排量。离心式压缩机是利用离心力压缩天然气。这种类型的压缩机利用叶轮进行工作。然后气体以高速排出，进入到扩散器中后速度降低，随后它的动能转化为静压。与往复式离心机不同的是，离心式压缩机不受限制且没有物理的挤压。

具有相对无限制的通道和连续流的离心式压缩机是内在高容量、低压比的机器，这样就很容易地适应了站内的一系列设置。这样的话，每台压缩机就只需要满足站内压缩率。通常，它的体积大于 100,000cfm，排出压力高达 0.69MPa。

旋转鼓风机被建成由内置一个或多个向相反的方向旋转叶轮的外壳。旋转鼓风机主要用于吸入和排出压力差少于 0.103MPa 的分配系统中。它们也常用于制冷和再生。旋转鼓风机有很多优点：大量的低压气体可以在相对较低的马力下处理；初始花费和维护费用都比较低；它安装简单，操作和维护容易；它占用最小的地面空间，几乎没有脉动流。它的缺点是不能承受高压，而且由于齿轮和叶轮的转动以及其叶轮和套管之间的不合理密封而使其运行时有噪声。如果运行时超过了安全压力，也会造成它的过热现象。通常，旋转鼓风机可输送高达 17,000cfm 体积的气流量，并可达到最大吸入压力 0.07MPa，而实现 0.07MPa 的压差。

选择压缩机时，需要考虑其压力—容积特性及其驱动器的类型。小型的旋转式压缩机（叶片或叶轮类型）通常是电动机驱动。大体积的正容积压缩机运行速度低，并且通常由蒸汽或气体引擎驱动。它们可能经由减速齿轮，通过汽轮机或是电动马达来驱动。由汽轮机或电动马达驱动的往复式压缩机作为常规高速的压缩机器，在天然气工业领域应用最为广泛。压缩机的选择需要考虑容积气体的供应能力、压力、压缩率和马力等因素。

燃气管道附属设备

一、听力部分

排水器

在正常水封高度气体不能排出的情况下，排水器用来排出在管网中连续产生的冷凝水。因此，排水器是必需的安全附属设施。通常，架空的管道排水器可以构建垂直和水平的排水器。为了保持排水器的有效高度，并且做到易于操作和维护，通常会有单一水密封和双重水密封两种形式。

闸阀应该安装在天然气管道的排水器上，与收集漏斗状的排水管垂直连接，排水器是切实可行的，但是当基础下沉时，它会增加部分管线的负担。排水器可以设置在户外，但是在寒冷地区必须要做好防冻措施。当设置在室内时，应当有好的通风条件。

1)discharge 2)overhead 3)effective 4)seal 5)feasible 6)burden 7)ventilation

二、口语部分

有用的句子

（1）无论在哪里使用蒸汽，必须安装疏水器，并且应使用自动阀门。

（2）疏水器的基本作用就是允许冷凝物流动，同时阻止蒸汽通过，直到它重新冷凝成水而失去热量。

（3）阀门是一种控制流体的流动和系统或进程中的压力的机械装置。

（4）众多的阀门种类和设计应安全适应各种各样的工业应用。

（5）阀体，有时称为外壳，是阀门主要的承压边界。

（6）阀体通常被浇注锻造成各种形状。

（7）阀体开口处的覆盖物称为阀盖。

（8）阀门的内部元素统称为阀门的内饰。

（9）对于有阀盖的阀门，这个阀盖就是第三个主要的压力边界。

（10）座封或者密封圈给阀盖提供了阀座表面。

（11）连接执行机构和阀盖之间的阀杆是用来定位阀盖位置的。

（12）执行机构操纵着阀杆和阀盖组合。

（13）执行机构可能是一个手动的操作手轮、手动的操作杆、马达操作器、螺旋管操作器、气动操作器或者液压油缸。

（14）大多数阀门使用某种填充料来阻止阀杆和阀盖之间空隙的渗漏。

（15）由于阀门必须在多种系统、流体介质和环境情况下工作，大量的阀门类型已经开发出来。

三、阅读部分

阀门

阀门是通过管线和清洗系统隔离或调节气体、液体和泥浆流速的一种装置。操作阀门所需的力可以通过手动或机械的方式来实现。阀门的机械附件（执行器）通常是电动或气动驱动。执行器可以手动控制（例如技术员按钮或开关），或是通过工厂控制系统来控制。可用的不同类型的阀门有数百种。

球阀：球阀，顾名思义，在球的中心钻了一个洞，旋转安装在阀体内。当球中洞的方向同管道的方向一致时，将会导致满流量流动。随着球中洞的方向慢慢地偏移管道的方向，流量将会受到限制，当洞的方向与管道的方向成九十度时，最终流量就会被完全阻断。注意球中洞的直径要小于管道的公称直径。

蝶阀：蝶阀和球阀的原理相似。但是，它是一个圆形的阀瓣而不是一个安装在阀体内的球（之所以称之为蝴蝶片是因为垂直轴周围的两个半圆看起来像是翅膀）。此外，蝴蝶片的方向决定了流量。当蝴蝶片的方向与管道的方向一致时（例如来液方向的横截面积最小），将会导致完全流通。随着蝴蝶片的方向逐渐偏离管道的方向，流量将会因增加的阻碍流体的面积而受到限制，最终当蝴蝶片的方向和管子的方向成九十度时，流量将会被完全阻断。

闸刀阀：闸刀阀，通常称为闸阀，被用作隔离阀。原理就仅仅是一把刀或是门掉落在液流之前。闸刀阀不应该用于限流（例如半开状态），因为闸刀的基础会迅速磨损，并且当关闭的时候密封不完全。闸刀阀的规格全，并且可以有手动滚轮或是气动执行器来提升或降下闸刀。

隔膜阀：隔膜阀（也被称为桑德斯，是在一个受欢迎的品牌）的工作原理是一个橡胶隔膜或者囊状物的开闭过程。桑德斯阀门非常适合限制或流量控制任务（例如半闭阀门来降低流量）。有两种主要的隔膜阀类型：屋脊式隔膜阀和直通式隔膜阀。隔膜的运动可通过手动或是气动执行器来控制（阀体的基础是相同的）。

止回阀（单向阀）：止回阀或单向阀的设计是确保流体只向一个方向流动。他们通常被用于水管线系统，并且直接安装在泵后面。最常见的止回阀就是阀片类型的（水平的或是垂直的）。当流量充足，阀片就被冲开；当流量减少（或是当泵失效而反流），阀片就会重新回到座封内，阻隔液流。

夹管阀：夹管阀用于流量控制应用，通常用来调节类似钻井液液面或增稠剂流固体的其他参数。阀门的操作原理是依靠于在压紧的两个活动杆（像老虎钳一样）之间变平的管子的挠性段。夹管越紧，流量越低。夹管机制可以是手动的，但是通常是气动操作，并且由 PLC 系统进行控制。橡胶段会因使用时间损坏，所以需要定期更换。最常见的夹管阀品牌是拉罗克斯。

泄压阀：顾名思义，泄压阀是当系统压力（例如在一个容器或管道里）太高时用来释放压力的一种安全装置，如果没有及时泄压，可能就会造成设备损坏并危及人员安全。最常见的类型就是用弹簧来操作的阀门。弹簧压力控制下的卸载阀座封在阀体内部，并且暴露于系统压力中。当系统压力高于弹簧压力，卸载阀将会移入座封，产生通向大气的出口，让系统气或是液体排出。

四、翻译部分

TPR 阀门和放散管

温度/压力释放阀或 TPR 阀是像锅炉和家用水加热器一样安装在水加热设备上的安全装置。TPR 阀设计为在水罐内的温度或压力超过安全值时，就会自动排水。

如果温度传感器和安全装置（比如 TPR 阀）失效了，系统内的水就会过热（超过沸点）。一旦罐破裂，水就会暴露于大气中，立即扩散成水蒸气，并且占据大约原始体积的 1,600 倍。这个过程可能推动加热罐像火箭一样穿过多层建筑，导致人员受伤和大量的财产损失。

很少出现水加热设备的爆炸，因为需要具备发生非常状况且冗余安全组件同时失效的事实。这些状况仅仅会由极端的疏忽和过期或失效的设备的使用造成。

如果水温（以华氏摄氏度测量）或压力[以磅每平方英寸（PSI）测量]超过了安全值，TPR阀就会启动。此阀应当与水加热罐长度相对应的放散管（也叫排出管线）相连接。此管道负责为从TPR排出的热水到达适当排放地点选择通道。

放散管满足下列要求是至关重要的。放散管应当：

（1）由合格的材料建成，如CPVC、铜、聚乙烯、镀锌钢、聚丙烯或不锈钢。PVC或其他容易熔化的不合格塑料不能使用。

（2）不小于阀门出口直径（通常不小于3/4in）。

（3）不减少从阀门到气隙（排出点）的尺寸。

（4）为避免阀门上过度的压力，要尽可能短和直。

（5）安装时要方便重力流排出。

（6）不被困住，因为积水可能会导致污染，并且回流到饮用水中。

（7）排到放地漏、间接的废水接收器或户外。

（8）不能直接连接到排水系统中，以免潜在污染饮用水回流。

（9）通过与水加热设备在同一房间的可见气隙排出。

（10）由于覆冰水可能堵塞管道，当要排出到可能结冻的户外区域时，首先输送到间接的废水接收器中（如通过位于加热区域气隙的水桶）。

（11）不终止于超过地面或废水接收器6ft(152mm)的地方。

（12）排出时不能造成烫伤事故。

（13）排出时不能造成结构性伤害或财产损失。

（14）当排放设备指示出现问题时，要排放到居住者易于观察的端点，并且阻止未被注意到的终端限制。

（15）和其他设备的排出管线、热水器锅或到排出点的压力排出管线分开铺设。

（16）不能到处都设阀门。

（17）不能有三通接头。

（18）为了避免限制，不能在管线末端用螺纹连接。

泄漏和激活：功能正常的TPR阀门在全部激活后能够从排出管线喷射出强有力的热水，而且没有微渗漏。TPR阀泄漏则表明它需要更换。这种情况很少见，TPR阀激活，房主应当立即关水，并联系合格的水管工人寻求援助并进行维修。

尽管检查员不会亲自来检查，但他们还是应建议房主每月检查TPR阀门一次。检查员应当给房主示范怎样关闭主水源，并说明它位于家中的主要供水阀的位置，或者位于安装有TPR阀门的设备水源关闭的位置。

任务 10 居民用户燃气应用设备

一、听力部分

燃气灶

在菲律宾，液化石油气(LPG)是燃气灶的常规燃料来源之一。LPG，作为燃料其使用在城市和郊区都是常见的，尤其是在供应地很容易获得。LPG 被广泛用于家庭使用的主要原因是：便于操作、易于控制，并且由于在烹饪过程中发出清洁的蓝色火焰。然而，由于在国际市场上原油价格持续上涨，LPG 燃料的价格已显著上涨，并且快速持续增长。目前，常用于普通住户进行烹饪的 11 千克 LPG 的花费与城市地区每罐 P540 的花费一样高，在郊区的某些地方甚至更高。对于典型的有四个孩子的家庭，一罐 LPG 可能在 20 到 30 天内就用完了，这仅仅取决于烹饪食物的数量。由于 LPG 燃料价格的问题，研究中心和机构面临的挑战是开发一种使用 LPG 之外的可利用替代能源进行烹饪的技术。作为可替换 LPG 的替代燃料，生物能是大有前途的。

1) common 2) convenient 3) cooking 4) continuously
5) typical 6) amount 7) alternative

二、口语部分

有用的句子

(1) 在大多数国家，来自管道的家庭用气通常是天然气的同义词。
(2) 到 20 世纪 20 年代带顶烧炉和内炉的煤气炉被多数用户广泛使用。
(3) 煤气炉的发展受限，直到送气到户的输气管线普及为止。
(4) 煤气炉的引进对于那些在家里做重体力活的家庭主妇来说是一个很大的慰藉。
(5) 天然气汽车主要是 CNG 汽车，这种现象近期不会改变。
(6) 汽车工业和整个社会在发展节能、环保汽车上已经达成共识。
(7) 当前在中国使用的 CNG 汽车主要是双燃料汽车（CNG 和柴油）。
(8) 最近五年 CNG 汽车发展迅速。

（9）在中国 CNG 站不足以提供充足的 CNG。

（10）与柴油汽车相比，天然气汽车的制造费用较高。

（11）由于高温这一因素，天然气汽车发动机要求零部件有高的可靠性。

（12）天然气主要由甲烷组成，并含有少量在炼化过程中没有除去的其他烃类。

（13）天然气的使用在减少重型汽车的 CO_2 排放方面是非常重要的，因为像充电那样的其他选择有局限性。

（14）现在，针对天然气的汽车及燃料技术已唾手可及，并且价格相对实惠。

（15）总的来说，与目前柴油汽车的污染排放相比，NGV 的污染排放量表现优异，尤其是在重型汽车方面。

三、阅读部分

气体加热器

最初的气体加热器利用了与前些年发明的本生灯的相同原理。它最早是在 1856 年由英国佩蒂斯和史密斯公司进行商业化生产。火焰局部加热空气。然后热空气通过对流扩散，从而加热整个房间。今天，同样的原理也应用到了户外露台取暖器或是相当于巨型本生灯的"蘑菇加热器"中。从 1881 年开始，燃烧器的火焰是用来加热石棉网，这是由英国工程师 Sigismund Leoni 设计的专利。后来，因为耐火土更容易铸造，所以它取代了石棉。现代气体加热器虽然使用了其他耐火材料，但是仍然以原来的方式工作。现代气体加热器已经有了进一步的发展，包括使用辐射热技术的装置，而不是本生灯的原理。这种技术形式并不是通过对流传播，更确切地说，是使热量被在传播路径上的人和物体吸收。这个加热形式对户外加热尤其有用，它比加热能自由移动的空气更经济。

烟道加热器：烟道加热器通常都是永久性安装。烟道，如果被恰当地安装在正确的整体高度，能够吸取大部分的热量释放。正确操作烟道气体加热器通常是安全的。

无烟道加热器：无烟道加热器又称为没有排气口的加热器、直排式加热器或无烟道点火式，可以是永久性安装或是便携式的，并且有时候还会安装有催化转换器。如果不遵守适当的安全程序，无烟道加热器是危险的。必须有足够的通风条件，但是通风就会降低房间温度，抵消了加热的效果。通风设备必须保持清洁，并且要在睡觉前关闭。如果操作正确的话，无烟道气体加热器的主要排放物质就是水蒸气和二氧化碳。如果燃烧不充分，类似一氧化碳和二氧化氮的有毒物质就会形成。

操作运行过程：家用气体加热利用机械或是电动的调温器来控制循环，使用阀门来驱动气流，使用电炉丝或引火灯点火。火焰加热在空气管中的散热器，但是在外部烟道，对流或风扇可以分散热量。

四、翻译部分

天然气汽车

　　天然气汽车（NGV）是一种替代燃料汽车，它使用压缩天然气（CNG）或液化天然气（LNG）作为比其他化石燃料更加清洁的替代燃料。天然气汽车不应与组成完全不同的以LPG为动力的汽车相混淆。在世界范围内，到2011年共有1480万天然气汽车，数量领先的几个国家中伊朗有286万辆，巴基斯坦有285万辆，阿根廷有207万辆，巴西有170万辆，印度有110万辆。亚太地区有680万辆NGV，居世界之首，其次是拥有420万辆的拉丁美洲。在拉丁美洲地区，几乎90%的NGV都是双燃料引擎，既能够使用汽油又能使用CNG。在巴基斯坦，几乎每一辆汽车都转为（或已制成）可替代燃料使用的，但特别保留了使用常规汽油运行的能力。

　　截至2009年，美国有114270辆压缩天然气（CNG）汽车的车队（大多数是公共汽车）；147030辆液化石油气（LPG）作燃料的汽车；3,176辆液化天然气（LNG）做燃料的汽车。其他普及以天然气为动力的公共汽车的国家包括印度、澳大利亚、阿根廷和德国。在经济合作与发展组织（OECD）国家大约有500000辆CNG汽车。巴基斯坦NGV汽车的市场份额在2010年为61.1%，紧随其后是美国的32%和玻利维亚的20%。2010年全世界NGV加气站的数量也增加到18202个，比2009年增长10.2%。现有的以汽油为燃料的汽车能够转换为以CNG或LNG为动力，并且能够专用燃料（只用天然气运行）或是双重燃料（用汽油或天然气运行）。重型卡车和汽车的柴油发动机也能够被转换，并且在另加包含火花点火系统的新的头部时可以用专用燃料，或者可以在柴油和天然气混合情况下运行，其主要原料是天然气和少量用作点火源的柴油。世界范围内，使用CNG为燃料的汽车数量逐渐增长。直到最近，在美国市场上本田思域GX是唯一可商购的NGV，但是现在福特、通用、Ram公司的汽车阵容中都拥有了双燃料汽车产品。福特的方法是让双燃料装备作为工厂选择，于是顾客可以选择一个授权合作伙伴来安装天然气设备。NGV加气站可位于任何存在天然气管线的地方。压缩机或液化工厂通常大规模建造，但是用CNG的小型家用加气站成为了可能。CNG也可与从垃圾填埋场或废水产生的不会增加大气中碳的浓度的沼气混合。尽管有它的优点，天然气汽车的使用面临着诸多限制，包括燃料的储存和用于燃料站递送和分布的可用的基础设施。CNG必须储存在高压缸内（操作压力为20.69~24.82MPa），并且LNG必须储存在低温绝热气瓶内（-260~-200°F）。这些气瓶比汽油或柴油罐占据了更多空间，汽油或柴油罐可以铸造成各种复杂形状来储存更多的燃料和利用更少的车载空间。CNG储罐通常放置在汽车的行李箱或机架上，这样就减少了其他货物的可用体积。这一问题可通过将罐安装在车体底部或顶部来解决，这样就释放了货物区域空间。正如其他替代燃料，广泛推行天然气汽车使用的其他障碍不仅是天然气的分布和燃料供应站，而且是低数量的压缩天然气和液化天然气加气站。

任务11 燃气管道施工安装

一、听力部分

美国能源局势快速多变,一桩油气管道交易抓住了契机

抽油机,海上钻井平台,炼油厂还有加油站,你看见的这些只是石油工业的冰山一角。把它们连在一起的是一个庞大且几乎隐形的地下管网。它价值不菲,正因如此ETP公司于4月30日表示,愿意出价53亿美元收购Sunoco公司。美国国内的石油和天然气产量正快速增长,ETP希望通过合并两家公司的管网,以输送更多油气产品。

ETP公司踌躇满志。去年其母公司ETE同意出价57亿美元,收购Southern Union公司及其天然气管网。最新的收购将使ETP成为全美第二大油气管道公司。Kinder Morgan公司并购El Paso之后排名第一,其交易将于今年晚些时候完成。

收购带来了Sunoco公司的储油设施、4900个加油站以及一家炼油厂的剩余资产。Sunoco打算把这些剩余资产剥离出来投入一家合资公司,该合资公司由ETP和私募股权公司Carlyle共同投资。但最诱人的部分是油气管道。ETP目前运营的油气管道干线达17500英里(约28160公里)。这些干线用于输送天然气以及丙烷、丁烷等液化天然气。加上Sunoco运输原油和精炼产品的6000英里管道,ETP对天然气输送业务的依赖程度将降低。交易完成后,ETP30%的收入将来自石油输送业务。

1)platforms   2) filling stations   3)invisible   4)output   5)ambitious
6)storage facilities   7)joint venture   8) pipelines   9)reliance   10) revenues

二、口语部分

1. 有用的句子

(1) 天然气由于密度低,不容易储藏和使用交通工具运输。天然气管线跨海也不可行。

(2) 天然气管道一般采用碳钢,根据管线的类型不同,直径在2~60in(51~1524mm)之间。天然气由压气站增压,混入硫醇类气体后才有气味。

（3）天然气比普通汽油便宜一两美元，而且燃烧后更清洁，这使它成为一个环境友好型的能源选择。

（4）西气东输工程是将天然气从西部输送到东部。

（5）管道敷设不一定是很复杂的工程，但必须小心施工，因为天然气可能会从连接不可靠的管线泄漏而引起爆炸。

2．对话

A：李先生，请问您有空吗？我有几个燃气管道的问题。不知道您能不能帮我解答一下。

B：好的。请说。

A：燃气管道可以埋入地下吗？

B：当然可以。比如用于城市管网建设的管道都是埋地的。但是在人口密集的地区，必须使用高质量的钢管。例如在西气东输的建设中，大部分的管线都是埋地的。一般情况下，埋地长输管线必须采用阴极保护。如今，所有用于城市管网的PE管都是埋地的。城市管网埋地是为了满足城市美化的要求。

A：如果管线埋地，应考虑哪些方面呢？

B：管线最好有一些坡度（特别是对于湿气），在最低点要设凝水缸。如果管线要穿过马路、河堤或者铁路，则需要装保护套管。严格意义上说，套管和PE管线之间应设检测管防止泄漏。要设阀门井和放散管，否则管道运行时，空气不能排放出来。如果可能的话，充氮也是一个选择，这样的话只需在管线末端设一个阀门，但最好是设阀门井。

A：埋地敷设和架空敷设这两种方式哪一种更经济呢？

B：我觉得不是哪种更合算的问题而是哪种更可行的问题。个人认为，架空敷设时，管道腐蚀很快，这会大大增加维护成本。如果可能话，强烈推荐使用埋地敷设，特别是在城市中压管线的建设中。如果条件不允许，可以选择其他方法。

A：城市管网都使用哪些管材？

B：输送燃气的管材必须有很好的机械强度、抗腐蚀能力、抗震能力和密封性，且应易于连接。可以选择的管材有以下几种：

(1) 钢管可塑性好，可承受较大的压力，容易焊接。按照不同的制造方法，钢管可分为无缝钢管和焊接钢管。

(2) 铸铁管和钢管相比，有很好的抗腐蚀性，因此它们被广泛用于城市低压管网中。铸铁管可分成灰口铸铁管和球墨铸铁管。但铸铁管很脆，不容易焊接，不能承受大的压力，所以在重点地区和动载荷大的地区，还是应采用钢管。

(3) 塑料管，硬聚氯乙烯管道和聚乙烯塑料管道比较轻，其抗腐蚀力强、摩擦阻力低、接口严密、拉伸力好、施工简单，因此被广泛用于中低压管网。但是它的老化问题还有待进一步研究解决。

A：谢谢您。您的解释对我的帮助很大。

三、阅读部分

天然气管道类型

黑色金属管道已不再是唯一的天然气输送材料。许多市政当局允许施工人员安装其他可替代的天然气管道，如轻量级的金属和塑料材质的管道。每种类型的天然气管道管有其独特的优势、劣势和成本。一旦你了解了常见类型的天然气管道的管材，您可以选择一个与你的预算和项目设计相匹配的材料。

不锈钢波纹管

不锈钢波纹管出现在大多数家庭中作为连接刚性供气线路和电器之间的挠性管。一些市政当局允许它被使用于整个供应系统。一段不锈钢波纹管的两端装有管道配件，可连接到电器或刚性管配件上。大多数市场上可买到的 CSST 有黄色塑料涂层或自然金属外观。CSST 的主要优势是它的灵活性，CSST 很容易扭曲来穿过障碍和穿墙而过。CSST 的主要缺点是成本高，一段 4ft 长的 CSST 成本超过 15 美元。

高密度聚乙烯

高密度聚乙烯（HDPE）是一种塑料材料，可以用于住宅天然气管道。HDPE 天然气管道共有的特点是灵活，水管工可从一卷中拖出一定长度的管线，简单地穿梭安装在建筑物的内部凹槽里。这种灵活性大大减少了安装时间。相比传统的铁管必须在两端使用螺纹装置，HDPE 管道配件可通过压紧连接到两端。相较传统铁管，HPDE 较高的价格也是普通住户使用它的障碍。然而，作为一种塑料材料，HDPE 几乎是对腐蚀免疫的，而钢铁最终会生锈。

黑铁管

传统的天然气管道管材黑铁管，可安全地将天然气包裹在厚厚的金属管道内。意外的钉子的冲击不会轻易击穿黑铁管，极大的力量才能弯曲或折断管子。普遍性和充足的供应使黑铁管成为最便宜的天然气管道类型。然而，尽管很容易找到水管工，安装黑铁管很费时间。此外，黑铁管暴露于水汽中会生锈和腐蚀。

四、翻译部分

（1）所有地下管道的建设应严格按照设计图纸进行。在耦合和连接配件的过程中，在焊接点必须要使用隔板。

（2）通过热处理测试的管道应当远离火焰或电弧。在管道组现场焊接过程中，应使用蜡笔或绘画笔标记图纸编号和密封焊接接头编号，以便于执行随机的无损检测和安装。

（3）如果发现图纸上没有标示的排气管、排水管或倾斜管道时，要求工人必须立即向技术人员报告。

（4）大直径的铸铁阀门和法兰阀门，存在由于法兰表面变形和螺栓不均匀紧密性而造成法兰损坏的风险。所以连接阀门时必须检查法兰表面，以确保安装之前没有出现因焊接不当造成的变形。

（5）应当特别注意与设备连接的管线。

（6）There must be proper temporary support for avoiding causing equipment dysfunction, excess deformation, even damage resulting from enduring resistant stress, bending stress and vertical stress.

（7）The welding slag on weld surface must be cleared, and the appearance quality of the weld surface should be inspected to ascertain whether there are welding defects such as pore, crackle and impurity. If there is any flaw, it must be repaired in time and repair record must be made.

（8）Form, material and installation location of pipe supports and hangers should be correct with required quantity. Their firm degree and welding quality should be acceptable.

（9）Weld and other parts to be inspected should be visually available.

（10）The welder should hold the qualification certificate of welding of boiler pressure vessel issued by the Labor Department before making operation.

燃气管道质量监督检验

一、听力部分

欧洲天然气价格之战

没人乐意收到天然气账单，而欧洲最大的天然气公司对于向天然气供应商支付的账单尤其感到沮丧，特别是俄国政府撑腰的巨头俄罗斯天然气工业股份公司的账单。有些燃气公司想削减开支，但是几乎没有任何制衡供应商的手段，所以成功的希望渺茫。欧洲的天然气定价是有几十年历史的长期合同方式，合同价格和石油挂钩。但是在市场相对自由的英国，天然气以供需决定的市场价格进行交易。在2008年天然气价格突然下跌时，在英国现货市场交易的小型欧洲天然气公司便拥有优势，而其规模更大的对手便受到损失。他们签署了一系列"照付不议"的合同，必须以大大高于现货市场的价格购买固定数量的天然气，而很大一部分根本就卖不出去。现货市场价格和合同价格的差异仍然存在。德国公司受俄国天然气管道的牵制尤为严重，所处地位非常不利。但是供应欧洲将近一半天然气的俄国和挪威对于满腹牢骚的客户显示了一定的灵活性。2009年底，和油价挂钩的天然气价格比现货市场高出50%，欧洲天然气买家纷纷请求供应商网开一面。作为回应，挪威国家石油公司允许在未来三年内部分采用现货价格(约占油价挂钩合同的25%左右)。俄罗斯天然气工业股份公司允许的比例更低一些，为15%至20%。

1)utilities 2)state-backed 3)slender 4)relatively 5)rivals 6)obliged
7)hostage 8)disadvantage 9)flexibility 10)generous

二、口语部分

1．有用的句子

（1）作业许可是通过书面形式授权可以进行某项作业的安全控制体系，从而使人员伤害和财产损失的风险降到最小。

（2）尽管许可证的格式和样式可能各不相同，但通常只包含三项主要内容。

（3）作业结束时，每个许可证必须验收签字并交回签发地。施工地点要打扫干净，所有工具、设备、垃圾等必须从施工区移除。一般规定要求许可证要存档一年。

（4）每个脚手架的设计和建造都要确保它可以承受必需的负载，其中包括人员、材料、工具、举升设备和脚手架的自重。

（5）脚手架在用于作业前必须先获得合格证。

（6）在切割或焊接作业过程中，压缩气瓶必须远离作业区。

2．对话

A：早上好。我们今天要检修管道吗？

B：是的。你把工具准备好了吗？

A：准备好了。为什么检修工作这么频繁？

B：因为燃气生产设备长期在高温、高压或深冷工艺条件下工作，很容易被腐蚀和磨损。所以燃气设备、管线、阀门和仪表不可避免地出现各种各样的缺陷。如果不能及时检测出这些缺陷，并采取相应的技术措施消除它们，设备会出现变形、断裂和穿孔等现象，这会导致严重的火灾。为了保证正常生产和防范事故，我们必须加强检测、保养和维修。

A：我明白了。对检修人员有什么特别要求吗？

B：是的。没有受过培训的检修人员不允许上岗。若没有获得工作许可证，任何人都不允许进入施工现场进行作业。所有的工作人员在作业前必须接受安全教育并通过考试。

A：工作许可包含哪些内容？

B：工作许可一般包括以下内容：检修作业项目、地点、起止时间、检修方法、潜在风险、安全技术措施、检修单位和现场负责人、安全监护人和检修人等。

A：今天我们有动火作业吗？

B：是的。请把这个标志立在那边。我们备有足够的灭火器吗？

A：是的。我们什么时候做可燃气体分析？

B：我们有些准备工作要做。首先我们要把点火区内的可燃物移除。

三、阅读部分

城市天然气门站的验收

站内管线施工完毕并经系统试压合格后,在投入运营前应对其进行验收。一般是由施工单位、生产单位和设计单位共同对站内工艺管道进行检查和验收。验收的具体要求如下:

(1) 城市天然气门站的安装工艺应符合设计和生产要求,流程合理,安装正确,设备稳固牢靠,便于操作和维护保养。

(2) 管线进行吹扫时应另设吹扫口,不准许脏物进入门站设备。特殊情况下,吹扫后应对设备、阀门重新清洗。

(3) 计量装置的安装应符合计量规程。

(4) 门站内所安装的各种仪表必须是经过校验、有出厂合格证的合格产品。无论是就地安装,还是墙上安装或表板安装,必须保证仪表平正,工作时不得有振动现象。

(5) 选择合适的调压阀,应灵敏、可靠、密封不漏,门站内汇管、分离器、收发球筒和各类阀门应通过单体强度测试和严密性试压,并且全站整体设备试压合格。

(6) 门站内阀门应操作灵活、性能可靠。站场设备无锈、漏、滴现象。加热装置(锅炉、水套炉)应正规安装、试压合格。一切都应在试运转中做到满足生产和设计要求。

(7) 绝缘法兰不泄漏、效果良好。

(8) 在门站有一套完整的安全装置(防雷、防火、防爆)以及环保设施。站内工艺管线工程竣工后,施工单位必须提供以下技术资料:站内工艺管线施工竣工图设计修改通知单、联络单、隐蔽工程检查验收记录、管线试压记录、焊缝探伤汇总表、阀门检查试压记录、原材料合格证等。当工程验收完毕后,就可由施工单位整理装订成册,交给建设单位签字后存档。

四、翻译部分

(1) 防腐层性能的检测包括黏结力(剥离强度)检测、防腐层缺陷检测和防腐层绝缘电阻测量。

(2) 阴极保护效果的检测包括土壤腐蚀性调查、保护电位测量、电流测量和绝缘法兰绝缘性能检测。

(3) Nevertheless, the anticorrosion of the buried metal pipeline in gas transmission station hasn't been paid enough attention always, to which only the coating is added and this has caused the serious corrosion consequence.

(4) Gas Transmission Station is the hub of natural gas pipeline system. The safe and efficient operation of the equipment in the station is the key to guarantee gas transmission.

(5) Once the piping system passes the pressure test, the system purging should be carried

out prior to the gas leakage test.

(6) The purging pressure should not exceed the design pressure of the container and the piping system.

一、听力部分

门站的运行管理

门站是长输管线终点配气站，一般设在长输管道的末端，也是城市燃气管网的起点处。它是城镇燃气输配系统的重要设施，也是输配气系统的起点和总枢纽。门站的工作人员应严格执行重点要害部位的安全保卫工作和消防安全工作规定。门站的工作人员应严格执行巡检、值班及交接班制度。同时，工作人员必须熟悉各种工艺流程及所使用设备的检测、安装、运行、维护、维修等工作，必须按照设备操作规程、维护检修规程和安全技术规程进行操作。门站应配备运行、维护和抢修的专职安全管理人员，并应建立相应的安全目标责任制度、安全生产管理制度、安全生产岗位操作责任制度及紧急事故抢修预案。门站的运行管理应符合国家现行有关强制性标准的规定。

1)terminal 2)located 3)implement 4)be familiar with 5)processes
6)procedures 7)full-time 8)responsibility 9)emergency 10)consistent

二、口语部分

1．有用的句子

（1）净化装置包括过滤器、除尘器和分离器等，一般在计量、调压装置之前设置于进站总管之上。

（2）门站的计量装置设置在调压装置前，用于燃气贸易计量。

（3）加臭装置可设置在门站的进口或出口处。

（4）控制系统的控制对象主要是进站和出站管道上设置的可远程操控的阀门。

（5）燃气输配系统的运行管理包括燃气管道及其附件、门站、储配站、调压站、调压柜（箱）、用户设施和用气设备。

（6）门站值班人员应按规定时间，定时巡查门站生产区。

（7）门站值班人员应保障门站的计量仪表、调压设备、压力容器、阀门、工艺管道及加臭装置的正常运行。

（8）门站的生产运行工艺参数应符合燃气公司制定的生产工艺要求或按照燃气公司生产调度部门的指示进行设定。

2．对话

A：杨经理，下午好！

B：下午好！很高兴见到您！

A：我恐怕要占用您的一些时间。请允许我向您进行自我介绍，我叫王明，毕业于中国石油大学，刚刚来本燃气公司工作了三个月。我有一些问题要向您请教。

B：好的，没问题。

A：您能向我介绍一下燃气储配站的相关知识吗？

B：好的。燃气储配站是城市燃气输配系统中储存和分配天然气的设施。它的主要任务是对天然气进行储存、调压，并向城市燃气输配管网分配燃气。

A：我明白了，您能告诉我燃气储配站是由什么组成的吗？

B：燃气储配站一般由储气罐、计量间、变电室、配电间、控制室、水泵房、消防水池以及生产和生活辅助设施等组成。

A：好的。您能向我简要地解释一下燃气储配站的作用吗？

B：当然可以了。首先，储存一定量的燃气以供用气高峰时调峰用。其次，当输气设施发生暂时故障或维修管道时，保证一定程度的供气。第三，对使用的多种燃气进行混合，使其组分均匀。第四，将燃气加压（减压）以保证输配管网或用户燃具前燃气有足够的压力。

A：非常感谢您，今天我收获很大。

B：不用客气。

三、阅读部分

天然气储配站的运行管理

储配站是城镇燃气输配系统的重要基础设施。它的主要功能是接受气源来气，并对燃气进行除尘、净化、储存、调压、计量、分配、气质检测，加臭后送入城镇或工业区的管网。储配站的主要设备有储气设备、过滤净化装置、计量装置、调压装置、测量仪表、气质检测设备、加臭装置、安全装置、加压设备、清管装置和监测与控制系统等。

储配站是安全防火的重点部位，工作人员应严格执行安全保卫和消防安全的规定。储配站的工作人员应严格执行巡检、值班及交接班制度。同时，工作人员必须熟悉各工艺流程及所使用设备的检测、安装、运行、维护、维修等工作，必须按照设备操作规程、维护检修规程、安全技术规程进行操作。储配站内所使用设备的运行操作必须严格执行安全技术操作规程，不可任意变更或减少操作。在储配站储存和输送天然气过程中，各工艺流程及所使用的设备严禁超压、超温、超速、超负荷。储配站应配备运行、维护和抢修的专职安全管理人员，并应建立相应的安全目标责任制、安全生产管理制度、安全生产岗位操作责任制及紧急事故抢修预案。根据"谁主管，谁负责"的原则，储配站站长对本站的安全负责，生产班组主值班员对本班的安全管理工作负责。在班组实行安全标准化管理，使班组作业程序标准化、生产操作标准化、生产设备标准化、作业场所环境标准化、工具摆放标准化、防护用品标准化、安全标志标准化等。储配站的生产设施与安全设施，必须严格执行同时设计、同时验收和同时投入使用，以确保安全生产。储配站的运行管理应符合国家现行有关强制性标准的规定。

四、翻译部分

（1）The routing inspection contents of pressure regulating device should include the operating conditions of the regulator, filter, safety relief facilities, instruments, meters and other equipment to ensure no leakage and other abnormal situations.

（2）The high-medium regulator station is generally once maintenance every quarter, and the medium-low regulator station is once maintenance every six months.

（3）According to the classification of action principle, the commonly used regulator can usually be divided into the direct acting regulator and indirect acting regulator.

（4）After regulator station passes the qualified acceptance and pressure testing, we pass gas to the total inlet valve at the outdoor, and then carry on the conversion ventilation for regulator station.

（5）The start of the regulator should be carried out after regulator station passes the qualified conversion air.

（6）一般情况下，出口压力一经调定后，调压站便一直运行下去。

（7）调压器是调压站内的主要设备，它是由敏感元件、控制元件、执行机构和阀门组成的压力调节装置。

（8）在寒冷地区采暖期前应检查调压室的采暖状况或调压器的保温情况。

（9）新投入运行和保养修理后重新启用的调压器，必须经过调试并达到技术标准后方可投入运行。

（10）调压站通常是由调压器、阀门、过滤器、安全保护装置、旁通管及测量仪表等组成。

任务 14 在役燃气管道维护

一、听力部分

城市燃气管道的日常维护

城市燃气管道的日常维护是管道正常运行的保证。日常维护的主要工作一般分为两大类：一类是维护管道本身，即巡线（包括检漏、外漏处理）、阀室维护、管线位置探测；另一类是处理外界因素对管道的影响，如第三方轻微破坏和损毁、违章建筑的处理等。工作人员定期在城市燃气管道（包括聚乙烯管道）上进行巡线检查、泄漏检查和管道检查，是城市燃气管网日常维护的重要工作。巡线频率根据地区的分类、管道材质和工作压力等做出决定。同时，对城市燃气管道上安装的阀门、凝水缸和补偿器等附件也要进行严格的日常维护。

1)assurance 2)maintain 3)namely 4)deal with 5)third-party 6)regularly
7)which 8)frequency 9)Meanwhile 10)installed

二、口语部分

1. 有用的句子

(1) 一般对地下燃气管道巡查应在燃气管道安全保护范围内，包括土壤塌陷、滑坡、下沉、人工取土、堆积垃圾或重物、管道裸露、种植深根植物及搭建建（构）筑物等。

(2) 在巡查中若发现问题，工作人员应及时上报并采取有效的处理措施。

(3) 聚乙烯（PE）管道切不可穿越密闭的空间。

(4) 高压、次高压管道每年检查不得少于一次。

(5) 新通气的管道应在 24 小时之内检查一次，并应在通气后的第一周进行一次复查。

(6) 燃气管道防腐层发生损伤时，必须进行更换或修补。

(7) 在燃气管道设施的安全控制范围内实施爆破工程时，工作人员应对其采取安全保护措施。

(8) 架空敷设的燃气管道上应有防碰撞保护措施和警示标志，工作人员应定期对管道外表面进行防腐蚀情况检查和维护。

2．对话

A：张先生，早上好！

B：早上好！很高兴见到您！

A：这些是清管器吗？它们能清扫管道吗？

B：是的，没问题。清管器就是在管道里移动的以清扫和检查管道为目的的设备。

A：投入使用前的城市燃气管道也需要进行清管器清扫吗？

B：是的，对公称直径不小于100mm的钢质管道宜采用清管器进行清扫。

A：我明白了。您能告诉我在清管器清扫过程中我们应该注意哪些问题吗？

B：好的，在清管器清扫过程中管道直径必须是同一规格，不同管径的管道应断开分别进行清扫。同时，对影响清管器通过的管件、设施，在清管前应采取必要措施。

A：您能简要地向我解释一下清管器清扫合格的要求吗？

B：嗯，清管器清扫完成后，当目测排气无烟尘时，工作人员应在排气口设置白布或涂白漆木靶板检验，5min内靶上无铁锈、尘土等其他杂物为合格。否则，工作人员可采用气体再吹扫至合格。

A：非常感谢，今天我收获很大。

B：不用客气。

三、阅读部分

燃气管道的投运

燃气管道安装完毕后，工作人员应依次进行管道的外观检查、吹扫、强度试验、严密性试验、工程竣工验收和置换投产等。

外观检查：地上燃气管道的检查内容主要包括管道和设备的设计要求、外观检验、气密性试验及其他附属工程符合技术要求的情况。地下燃气管道的检查内容包括气密性、管基、坡度、覆土、深度、借转角度、管位、操作工艺等。

管道的吹扫：对球墨铸铁管道、聚乙烯管道、钢骨架聚乙烯复合管道和公称直径小于100mm或长度大于100m的钢质管道，可采用气体吹扫；对公称直径不小于100mm的钢质管道，宜采用清管器进行吹扫。管道吹扫介质宜采用压缩空气，严禁采用氧气和可燃性气体。

强度试验：在满足强度试验情况下，对设计压力大于0.8MPa的钢质管道，宜采用清洁水进行试验，其他情况使用压缩空气。工作人员应严格按照强度试验的步骤和要求进行，合格后应填写管道系统试验记录。

严密性试验：严密性试验应在强度试验合格、管道全线沟槽回填后进行，但管道的焊缝、接口等应检部位应留出，不予回填。待严密性试验合格之后，完成这些部位的防腐，再回填。严密性试验介质宜采用空气。所有未参加严密性试验的设备、仪表、管件，应在严密性试验合格后进行复位，然后按设计压力对系统升压，不漏为合格。严密性试验合格后，工作人员应填写管道系统试验记录。

工程竣工验收：工程竣工验收是检验工程质量必不可少的一道程序，也是保证工程质量的一项重要措施。验收程序必须规范、严格，以彻底排查工程中的质量、安全隐患。

置换投产：置换投产各小组应定员、定岗、定职责，必须落实到位。置换前应根据置换方案对所有参加置换人员进行技术交底、安全交底和应急预案说明。置换方式可以选用燃气直接置换，也可以选用惰性气体作为中间介质间接置换。一般对大型燃气输配系统采用间接置换，而对小型的或已运行系统接出的分支燃气管道多采用直接置换。工作人员应严格按照置换的步骤和要求进行。置换完成后，首先整定调压与安全控制装置，使系统升压至预定压力，再对系统进行全面检查。确认状态完好后，工作人员即可开始置换及向用户供气，经试运行后转入正常运行。

四、翻译部分

（1）For gas pipeline under the railway and road, we can check through inspection well or leak detection tube whether there is gas leakage.

（2）At present, the commonly used methods of leak detection in the world are: visual leak detection method, sound leak detection method, the sense of smell leak detection method and tracer leak detection method.

（3）For the leakage of the higher pressure gas pipeline on the ground, we can rapidly shut off the upstream and downstream valves of gas pipeline network to block leakage pipe section.

（4）When the concentration of the environment is within the concentration range of explosion and poisoning, the staff must conduct the forced ventilation, and can go on with their work after reducing the concentration.

（5）The rush repair of gas facility leakage is appropriately made after reducing gas pressure or cutting off gas source.

（6）当泄漏处已发生燃烧时，工作人员应先采取措施控制火势后再降压或切断气源，严禁出现负压。

（7）修复供气后，抢修人员应进行复查，确认不存在不安全因素后，方可撤离事故现场。

（8）在液化石油气泄漏抢修时，还应备有干粉灭火器等有效的消防器材。

（9）为了防止积水堵管，必须制定出严格的运行管理制度，定期排除集水井中的凝结水。

（10）清除杂质的办法是对干管进行分段机械清洗，一般按50米左右作为一个清洗管段。

任务 15 燃气场站安全管理

一、听力部分

燃气站场的类型及功能

城市燃气输配系统是指从接收长输管道供气的门站开始直至用户用具的整个系统，它包括门站、储配站等部分。城市燃气站场在城市燃气输配系统中有着十分重要的作用，一般起到储存、气化、调压、计量及加臭等作用。根据其作用，城市燃气站场包括门站、调压站、储配站、气化站、混气站、加气站等。随着场站的功能不同，那么工艺流程和设备设施也有所不同。我们可以将每个功能看作一个模块，根据实际需要进行不同模块之间的组合来实现需要的功能。

1)whole   2)receiving   3)and so on   4)generally   5)pressure regulating
6)According to   7)fuelling station   8)different   9)combination   10)required

二、口语部分

1．有用的句子

（1）LNG 气化站通常指具有接收 LNG、储存并气化外输功能的站场。

（2）LNG 气化站工艺设备主要有储罐、气化器、调压装置、计量装置和低温泵等。

（3）天然气门站是长输管线终点配气站，也是城市天然气接收站，具有净化、调压、计量、储存及加臭等功能。

（4）调压站的任务是将输气管线的压力调节至下一级管网或用户所需要的压力，并使调节后的压力保持稳定。

（5）液化石油气供应系统的站场类型主要包括储存站、灌装站、储配站、气化站、混气站和瓶装供应站等。

（6）液化天然气工业系统包括天然气的预处理、液化、储存、运输、接收站、再气化等。

（7）天然气储配站是城市燃气输配系统中储存和分配天然气的设施。它的主要任务是天然气的储存、调压，并向城市输配管网分配燃气。

（8）LNG 气化站主要作为输气管线达不到或采用长输管线不经济的中小型城镇的气源，另外也可作为城镇的调峰应急气源。

2．对话

A：杨教授，早上好！很高兴见到您！

B：早上好！很高兴见到您！

A：我恐怕要占用您的一些时间。请允许我向您进行自我介绍，我叫王林，毕业于东北石油大学，刚刚来本燃气公司工作了三个月。我有一些问题要向您请教。

B：好的，没问题。

A：您能告诉我燃气站场安全生产管理的方针是什么吗？

B：当然可以了。安全生产管理坚持"安全第一，预防为主"的方针。

A：谢谢您。如果燃气站场发生爆炸或起火事故的时候，我们应该怎样进行处理呢？

B：首先，现场工作人员应立即通知调度中心和上级领导，有人员伤亡时，在保证自身安全的前提下，现场工作人员应立即组织救助。其次，现场工作人员应迅速有效地切断上游来气和出站气，现场工作人员应迅速组织救火工作，火势较大时，立即通知119援助。

A：真心地感谢您，今天我收获很大。

B：不用客气。

三、阅读部分

燃气站场安全管理规定

城市燃气站场的安全管理涉及站区内的各个工艺装置及站内管道系统等多个方面，是燃气安全生产的重中之重。因此工作人员应严格执行以下安全管理规定：

（1）进站人员必须遵守站内各项管理规定。

（2）严禁在站区内吸烟及使用明火。

（3）严禁携带火种及易燃易爆物品进站。

（4）外来车辆及人员进入站区必须进行登记，出、入站时要主动接受检查。

（5）进入站区的非工作人员（参观人员除外），需持有本人有效证件或部门负责人批准后方可进站。

（6）参观人员需持有上级主管部门签发的介绍信或有公司领导陪同方可进站。

（7）机动车辆进入生产区需加戴防火帽，或进入站内熄火。

（8）严禁穿带钉鞋进站，进入生产区的人员应一律穿着防静电服装。

（9）进入站场生产区内人员必须关闭手机等通信设备。

（10）未经同意，禁止动用站内任何消防设施和工具。

（11）未经公司批准，站场内禁止拍照和录像。

（12）外来办事车辆必须按指定位置停放，不得随意停放。

(13) 进入站区内施工的车辆，在携带上级主管部门颁发的进站施工许可证、加戴防火帽后方可进入站区，且必须停放在施工规定区域内。

(14) 驾驶人员要严格执行进站人员管理规定，严禁在站区内随意走动及随车携带无关人员进入站区。

(15) 做好其他各项安全防范措施。

四、翻译部分

(1) The production and operation of city gate station must be strictly carried out in accordance with the operation management system that the company established.

(2) The process parameters of production and operation of city gate station should be consistent with the production process requirements that the company established or be set in accordance with the instructions of the company's production scheduling department.

(3) The production equipment and process pipeline of city gate station should be maintained in accordance with the relevant provisions of equipment management to keep in good condition.

(4) The pressure vessels and safety accessories of city gate station should be regularly inspected according to the provisions of the national quality and technical supervision department.

(5) The staff of city gate station should operate strictly according to the operating rules, the process requirements that the company established and the dispatching instructions.

(6) 门站的工作人员应具备在生产设备出现异常时进行紧急处置的能力。

(7) 接班人实地检查设备运行工况与交班人提供的情况不符或不清楚时，有权拒绝签字，并及时向领导报告。

(8) 作业场所安全设施齐全完好，安全警示标志齐全、醒目。

(9) 岗位操作人员上岗前必须接受安全技术培训，达到岗位安全要求并经考试合格后方可上岗。

(10) 门站的工作人员应定期对消防设施进行维护保养，并做好记录。

任务 16 燃气管道安全管理

一、听力部分

燃气管网安全管理的意义

在城市燃气输配系统中压力管道是生产、生活中广泛使用的可能引起燃爆或中毒事故的危险性较大的特种设备。燃气管道作为压力管道的一种，其特点为多埋于地下，经过人口密集区，施工与检验、检修难度大，受地理条件的限制及外界原因的影响，易发生火灾、爆炸或中毒事故，造成较大的社会影响及危害。因此，对燃气管道存在的安全问题进行分析，提出有效的防范措施，对确保燃气管网安全、稳定运行具有非常重要的现实意义。

1) special   2) higher   3) characteristics   4) underground   5) populated   6) geographical
7) result in   8) Therefore   9) put forward   10) significance

二、口语部分

1. 有用的句子

（1）只有严格按照燃气管道安全操作规程中规定的操作压力和操作温度运行，才能保证管道的使用安全。

（2）燃气管道在运行中应尽量避免压力和温度的大幅度波动，应尽量减少管道的开停次数。

（3）燃气管道操作人员必须经过安全技术和岗位操作的学习培训，经考试合格后才能上岗并独立进行操作。

（4）燃气管道操作人员必须熟悉本岗位燃气管道的技术特性、系统结构、工艺流程、工艺指标、可能发生的事故和应采取的措施。

（5）燃气管道工作人员在巡检过程中发现异常情况应及时汇报和处理。

（6）燃气管道工作人员应妥善保管各种巡视检查记录，以备查验。

（7）在对燃气管道维护保养过程中，工作人员应及时消除跑、冒、滴、漏。

(8) 燃气管道维护保养工作是延长燃气管道使用寿命的基础。

2．对话

A：李老师好！很高兴见到您！

B：张亮好！很高兴见到您！

A：我恐怕要占用您的一些时间。我有一些问题要向您请教。

B：好的，没问题。

A：您能否告诉我天然气管线的安全置换有哪几种方法？

B：当然可以了。天然气管线的置换方法有直接置换和间接置换两种。

A：谢谢您。您能给我解释一下吗？

B：首先说说直接置换法。该方法的工艺操作简单、便捷，比较适合一些管容量比较小，且压力级制比较低的天然气管道工程。只要新旧管道连通之后，就可以通过燃气压力使燃气新管道内的空气置换出去，在取样试验合格之后就能够投入使用了。该种置换方法就存在一定的危险性。因此应该采取相应的安全措施。其次，间接置换法。该方法往往应用在一些管容量比较大、压力级制比较高的天然气新建管道工程，先使用惰性气体将管道中的空气置换出去，然后再向其中输入燃气进行置换。该方法在置换过程中虽然安全性较高，但是费用也较直接置换法高出很多，且工序繁琐。

A：真心地感谢您，今天我收获很大。

B：不用客气。

三、阅读部分

燃气管道档案管理

整理好燃气管道的技术资料及做好档案的管理工作，可为管好、用好燃气管道奠定稳固的基础。在燃气管道管理的各项工作中，做好技术档案资料管理工作具有十分重要的意义。建立一套完整无误的管理档案体系，可以掌握燃气管道设计、制造、维修、检验、使用过程中遗留的质量问题。依靠完整准确的档案资料，可制定出合理科学的检验计划，有针对性地进行缺陷检验，以确定管道使用条件和期限，进一步加强燃气管道的全过程管理。可以说，档案资料的完善与否是衡量燃气管道管理工作水平的重要尺度之一。

燃气管道基础档案资料管理工作应高标准、严要求，并持之以恒、常抓不懈。主管部门和企业对档案的形式、规格、内容应提出原则要求，使用部门必须按统一标准具体执行。

燃气管道技术档案主要包括原始技术资料和使用情况两部分，构成了燃气管道从设计、制造、安装、使用、检验和修理改造直至报废的全过程的全部资料信息。

原始技术资料是指燃气管道在设计、制造和安装过程中的基本信息，它由设计、制造和施工安装的企业提供。燃气管道实际运行情况记录包括运行中的主要工艺参数，管道历次检验、修理、改造和变更等情况，还包括因历史原因造成管道、设备资料短缺、遗失或内容不详。应定期检验，至少应补充所缺的必要资料。重要燃气管道必须做到每一条管道都建立档案。

除了各类管道的技术档案以外，燃气管道档案管理还应收集和整理其他有关技术资料，包括各类手册、图册、标准、规范、规程、制度、重大检修方案、技术总结、下发的各种文件、各类报表及其他各类记录。

四、翻译部分

（1）The detection of buried gas steel pipeline is divided into the external and internal detection.

（2）The external detection by means of the combination of non excavation and excavation methods is the detection of the performance status of gas pipeline, including the corrosion environment investigation of gas pipeline, the detection of corrosion protection condition and the detection of pipeline body safety condition.

（3）The internal detection is the detection of the performance status of gas pipeline through the intelligent crawler of gas pipeline.

（4）Alternating current (AC) attenuation method can be used for the evaluation of the overall conditions of the external coating of gas pipeline and the location of the damaged points on the coating.

（5）Close interval potential survey method (CIPS) can measure the size of the damaged area on the coating with higher detection accuracy, and at the same time can record the cathodic protection state of the measured pipeline.

（6）直流电压梯度（DCVG）技术适用于带有阴极保护的埋地管线，具有较高的定位准确度和测量准确度，且不受周围平行管道的影响。

（7）直流电流电位法的原理是通过阴极电流测量电流衰减及电位偏移来计算防腐绝缘层的绝缘性能参数。

（8）由于城市燃气管道不同于长输管道，要想通过非开挖手段获得管体腐蚀状况信息，采用导波检测技术是重要选择项目之一。

（9）城市燃气管道内壁产生腐蚀等缺陷时，国外常采用被称为清管器的检测仪进行检测。

（10）管道内检测装置一般比较庞大、价格昂贵、使用费高，且专用性强。

任务 17 安全应急管理

一、听力部分

燃气管道故障处理

在燃气管道运行中，最易出现的问题是漏气与管道阻塞，及时发现泄露是预防、治理的前提，是燃气管网运行管理的主要任务之一。因此在燃气管道管理中，应重点抓住管道漏气与管道阻塞两个问题，以及由管道漏气与阻塞引起的故障处理。工作人员应按相关规程要求，按期进行压力管道的在线检验，及时发现问题，消除隐患，以保证管道畅通与安全运行。目前，世界上通用的泄露检测方法有视觉检漏法、声音检漏法、嗅觉检漏法和示踪剂检漏法等。

1)most likely   2)blockage   3)premise   4)focus on   5)as well as   6)on-line
7)according to   8)smooth   9)At present   10)smell

二、口语部分

1．有用的句子

（1）钻孔查漏就是沿着燃气管道的走向，在地面上每隔一定距离（一般2至6米）钻一孔眼，用嗅觉或检漏仪进行检查是否漏气。

（2）挖探坑就是在管道位置或接头位置上挖坑，露出管道或接头，检查是否漏气。

（3）观察植物生长检查漏气是一种经济有效的方法，因为经地下管道漏出的燃气扩散到土壤中将引起树木及植物的枝叶变黄和枯干。

（4）如果发现凝水缸抽水量突然大幅度增多，有可能是燃气管道产生缝隙，地下水渗入了凝水缸。由此也可以预测到燃气的泄漏。

（5）如果地下燃气铸铁管道上砂眼漏气，可采用局部钻孔堵塞的方法进行处理。

（6）如果地下燃气铸铁管产生裂纹或折断，可使用夹子套筒处理。

（7）在燃气管道出现漏气点时，尤其是较高压力燃气管道漏气，此时应迅速关断燃气管道上的阀门，以隔断漏气管段来限制事故扩大，并应立即采取措施，对漏气点进行紧急处理。

（8）由于钢管腐蚀穿孔、裂缝、折断等原因，漏气发生在管身时，为迅速清除事故，可采用急修管箍。

2．对话

A：王教授，早上好！我正在自学液化石油气储配站知识，但是我遇到了一些困难。我可以问您一些问题吗？

B：当然可以了！

A：谢谢您！我在书中看到了一个不熟悉的问题。那就是关于液化石油气的泄露。

B：嗯，液化石油气具有易燃易爆、易蒸发、热膨胀性等危险特性，在生产、运输、储存容量较大的储罐区储存过程中一旦泄漏，极易导致燃烧爆炸、中毒事故，甚至引发恶性环境污染事故。

A：我明白了。您能介绍一些液化石油气易泄漏部位方面的知识吗？

B：好的。首先是管道法兰、阀门等连接密封部位失效或泄漏，其次是管线泄漏，第三是罐底及罐底阀件腐蚀泄漏，第四是罐体泄漏。

A：针对液化石油气的泄露有哪些处理技术呢？

B：堵漏、倒罐、注水这三项液化气泄漏处理技术在液化石油气储罐泄漏事故中的应用意义重大，可有效避免火灾、爆炸及环境污染事故。这种技术措施可以用在其他化工燃料储存设备上，加以推广应用。

A：现在我对液化石油气的泄露有了一些了解。真是太感谢您了！

B：不用客气。

三、阅读部分

城镇燃气事故应急预案的主要内容

城镇燃气企业应根据国家有关法律法规的规定和所在地政府、燃气行政主管部门制定的燃气重大安全事故应急救援预案，结合本企业具体情况，制定燃气重大安全事故应急预案；健全抢险组织机构，成立专业应急抢险队伍，配备完善的抢修、抢险设备和交通通信工具，并定期组织演练，积极组织开展事故应急救援知识培训教育和宣传工作；出现燃气安全事故时及时向所在地燃气行政主管部门报告，并立即组织进行抢险。

应急预案按功能类别一般可分为四类，即综合应急预案、专项应急预案、现场应急预案和单项应急预案。事故应急预案的编写要具有科学性、实用性和权威性。《城镇燃气设施运行、维护和抢修安全技术规程》中规定：对城镇燃气设施抢修应制定应急预案，并应根据具体情况对应急预案及时进行调整和修订。应急预案应报有关部门备案，并定期进行演习，每年不得少于一次。

应急预案可包括下列主要内容：

(1) 基本情况；

(2) 危险目标及其危险特性和对周围的影响；

(3) 危险目标周围可利用的安全、消防、个体防护的设备、器材及其分布；
(4) 应急救援组织机构、组织人员和职责划分；
(5) 报警、通信联络方式；
(6) 事故发生后应采取的处理措施；
(7) 人员紧急疏散、撤离；
(8) 危险区的隔离；
(9) 检测、抢险、救援及控制措施；
(10) 受伤人员的现场救护、救治与医院救治；
(11) 现场保护；
(12) 应急救援保障；
(13) 预案分级响应条件；
(14) 事故应急预案终止程序；
(15) 应急培训和应急救援预案演练计划。

需要特别指出的是，一旦事故被识别并确认后，应急预案随即启动。只有在下述几方面工作完成之后才能确认事故应急救援工作的结束，即此次应急预案终止：

(1) 造成事故的各方面因素以及引发事故的危险因素和有害因素已经达到规定的安全条件，生产、生活恢复正常。

(2) 在事故处理过程中，为防止事故次生灾害的发生而关停的水、气、电力及交通管制等恢复正常。

事故应急救援结束后，经对现场进行检测，确认造成事故的各方面因素以及事故引发的危险因素和有害因素已经达到规定的安全条件，由事故应急领导小组下达终止事故应急预案的指令，通知相关部门及地方政府危险解除，由地方政府通知周边相关部门和地区民众。

实战（或演练）后应及时对应急预案的效果进行评价或评审，并及时对应急预案进行修改、完善。

四、翻译部分

(1) When the outlet pressure is too high, we should timely regulate the pressure of the regulator, and should pay attention to the adjusted pressure within the prescribed scope.

(2) When regulator failure occurs, we should timely close the inlet valve of the regulator, and should use the standby regulator at the same time.

(3) For the direct acting regulator, due to the rupture of the diaphragm, the gas can be made directly from the high-pressure side to the low pressure side to result in the transmission of the high-pressure natural gas, and the diaphragm must be immediately replaced.

(4) For the direct acting regulator, because the imprecisely closed valve on the valve port or the diaphragm leakage can cause the outlet pressure rising or shutoff pressure too high, at the moment, we should grind the valve port and replace the rubber gasket and the diaphragm.

(5) For the direct acting regulator, due to the outlet pipeline with water, there is obvious beating outlet pressure, at the moment, we should eliminate the water in the condensate drainage near the pipeline.

(6) 对于 T 型调压器，调压器皮膜破裂往往会造成调压器自动关闭停止供气。遇此情况，只需更换皮膜。

(7) 对于活塞式调压器，安装质量不好往往造成出口压力及调整压力的波动或跳动。因此必须严格掌握组装调压器的技术标准。

(8) 对于自力式调压器、调压器皮膜破损、阀口磨损、阀垫损坏，均会造成调压器关闭压力过高或关闭不严。此时，应更换皮膜及阀垫，研磨阀口。

(9) 对于自力式调压器，指挥器弹簧失效、喷嘴堵塞、主调压器阀口堵塞或阀垫因腐蚀而发胀，均会造成出口压力下降。此时，应疏通喷嘴，清洗调压器，更换阀垫及弹簧。

(10) 对于曲流式调压器，由于指挥器上阀口打不开，指挥器出口节流阀完全堵塞，造成调压器不能调节起动。此时，应重新装配指挥器，清洗节流阀。

任务 18 液化天然气

一、听力部分

石油价格下跌对各国的不同影响

自 6 月以来，世界原油价格下降了大约 30%，这帮助了世界上许多经济体。当人们在燃料方面的花费减少时，他们可以增加在其他商品上的支出。但是石油价格下降也意味着国家石油公司收益减少。为石油行业提供服务的私营石油生产商和公司也受到了影响。美国石油产量的增加是世界石油价格下跌的一个原因。美国开发的水力压裂技术使得石油产量增加。水力压裂技术是指向地下泵入水和化学物质，迫使天然气和石油流出。这种技术通常用于从页岩中提取化石燃料。

吉姆·克兰是赖斯大学詹姆斯·贝克公共政策研究所的能源专家。他表示，页岩中出产的石油改变了市场。

中国和欧洲经济发展速度放缓也减少了对石油的需求量。过去，石油输出国组织会通过减少产量来回应需求的减少。12 个成员国经常合作来影响全球石油价格。然而，吉姆·克兰表示，成员国也想保卫自己的市场份额，不愿意减少产量。"他们愿意看到价格稍微下降一点，看看是否能将一些生产成本较高的生产商挤出市场。"然而，这会影响美国的小型石油生产公司。价格下降给这些公司带来的边际收益影响更加严重。"如果你是一家小公司，正在考虑投资来提升产量，然而如果价格下降到每桶75美元，你就会三思。"受到价格下跌压力的不仅仅是美国的公司。石油出口占尼日利亚出口量的83%，占国民经济的70%。财政部长 Ngozi Okonjo-Iweala 最近宣布了政府增加收入的措施。

1)fuel 2)profits 3)affected 4)fracking 5)shale 6)decreasing 7)influence 8)sweat 9)investing 10)measures

二、口语部分

1．有用的句子

（1）再气化是将温度为零下162℃(260°F)的液化天然气再转化成大气温度下的气体状态的过程。

（2）在液态时，液化天然气不会爆炸，不能燃烧。

（3）液化天然气要燃烧，必须先蒸发，然后与空气按适当的比例混合（易燃范围是5%～15%），然后被点燃。

（4）在泄漏的情况下，液化天然气会迅速蒸发而转变成气体，并与空气混合。如果这种混合物在易燃范围内则有被点燃的风险，会造成火灾和热辐射危害。

（5）典型的天然气液化的过程就是气体首先被提取并被运送到加工厂进行净化，移除其中的水分、油、泥，并移除其他气体如二氧化碳和硫化氢的过程。

2．对话

A：你好史密斯先生，我能问您几个关于中国石油资源的问题吗？

B：当然，也许我不能回答您所有问题，但我会尽力。

A：谢谢。很荣幸能采访您。在中国都有哪些石油化工原料？

B：石油、煤、天然气是主要的化工原料。石油作为主要能源，对人类社会的发展有重要贡献。

A：我很同意您的观点。您能说说中国石油资源的情况吗？

B：可以。尽管中国富有石油和煤炭资源，但我们也面临着一些困难。

A：是什么困难？您能详细说一下吗？

B：首先，由于过度开采，石油资源面临枯竭。其次，石油的过度使用造成了严重的环境破坏和污染，特别是因二氧化碳的过量排放而引起的温室效应。

A：是的，这也是一个全球性的问题。

B：同时，石油资源与国家的经济和人民的生活息息相关，它是国家经济发展的重要产业。

A：面临着这些问题，中国将采取哪些措施？

B：一方面，我们尽最大努力减少污染，另一方面，我们将充分利用新科技。

A：我听说你们已开始开发新的石油天然气资源，并利用了新的技术，是吗？

B：是的！油气勘探的新领域和新技术是中国"十三五"规划期间是能源领域的重要的目标，也是今天的石油行业的焦点。

A：明白了，谢谢！

三、阅读部分

不断增长的天然气需求

天然气在美国的能源供应以及实现国家的经济和环境目标中发挥着至关重要的作用。虽然预计到2025年，北美天然气产量将逐步提高，但消费量已经开始超过可用的国内天然气供应。随着时间的推移，这种供需差距将会扩大。

增加液化天然气(LNG)的进口以确保美国消费者在未来获得充足的天然气供应是被提出的选择方案之一。液化技术使本有可能被"搁浅"的天然气进入主要市场。天然气资源丰富的发展中国家也很热衷于依靠天然气的液化出口把天然气换成货币。反过来说，更多的拥有很少或根本没有国内天然气资源的发达国家依赖进口。

液化天然气储存的努力始于1900年，但直到1959年，世界上第一个液化天然气船才把货物（LNG）从路易斯安那州运到到英国，证明了越洋液化天然气运输的可行性。日本在1969年首次从阿拉斯加进口液化天然气，并且由于在20世纪的70年代和80年代大量扩张液化天然气进口，而使它跻身液化天然气贸易的前沿。美国在20世纪的70年代首次从阿尔及利亚进口液化天然气。

阿尔及利亚是世界第二大液化天然气出口国。通过阿尔及利亚国家石油天然气公司的四个液化工厂，主要为欧洲(法国、比利时、西班牙和土耳其)和美国提供液化天然气。尼日利亚主要出口到土耳其、意大利、法国、葡萄牙和西班牙，但也在短期合同下向美国提供产品。特立尼达和多巴哥共和国出口液化天然气到美国、波多黎各、西班牙和多米尼加共和国。一个埃及工厂在2005年首次出口液化天然气，预计供应法国、意大利和美国。从2006年开始，挪威计划从Melkoya岛出口液化天然气到西班牙、法国和美国的市场。

四、翻译部分

（1）LNG is short for liquefied natural gas. LNG is natural gas (predominantly methane, CH_4) that has been converted to liquid form for ease of storage or transport.

（2）LNG is odorless, colorless, non-toxic and non-corrosive.

（3）The liquefaction process involves removal of certain components, such as dust, acid gases, helium, water and heavy hydrocarbons.

（4）A: Why does natural gas need to be processed before it can be used as a fuel?

B: In its "raw" state, it contains a lot of impurities, even water. After all those are gotten rid of, the gas can burn consistently and reliably. Water can cause formation of hydrate in pipeline and equipment, resulting in pipeline and valve choking. H_2S is extremely poisonous gas as well as an important source of sulfur in chemical industry. CO_2 will decrease the heat value of the natural gas, while the sour gas will cause the heavy corrosion of the tubing and surface equipment in the presence of water.

（5）液化天然气在运输过程中，被储存在特殊设计的带双壳的轮船中，以防止货物损坏或泄漏。有几个特殊的泄漏测试方法来测试一艘LNG船液舱薄膜的完善性。

（6）天然气运输和供应是天然气业务的一个重要方面，因为天然气产地通常非常远离消费市场。天然气在运输中，其体积远远超过石油，而且大部分气体通过管道运输。

（7）在前苏联、欧洲和北美之间有一个天然气管道网络。天然气即使在较高的压力下密度也较低。通过高压管道输送的天然气的速度比石油快得多，但由于天然气密度低，每天传递的能量只是石油的五分之一。

（8）在管道末端，天然气在被航运前通常被加工成LNG。

（9）用于将产品从液化天然气船转至岸边存储设施的短的液化天然气管道已建成。

（10）更长的管道也正在建设中，它可以允许液货船在距离港口设施更远的地方卸载LNG。

任务 19 压缩天然气

一、听力部分

需要被关注的清洁能源

天然气生产蒸蒸日上，但它的清洁能源形象却受到质疑。

美国的天然气开采呈现这样一副蒸蒸日上的景象。随着钻井技术，特别是充满争议的水力压裂法（又称破裂法）的广泛采用，已探明的天然气储量不断增加。2009年末，美国估计其天然气储量为283.9万亿立方英尺（合8万亿立方米），比上一年增长了11%。2010年美国开采的天然气为22.6万亿立方英尺，而2005年为18.9万亿立方英尺。同期天然气的出井价也从每千立方英尺7.33美元下降到4.16美元。

但有人怀疑天然气是否真得就那么清洁环保。一方面，水力压裂需要耗用大量的水，水作为一种陆地资源，价值被严重低估。另一方面，这种方法还会释放大量的极易吸热的甲烷气体。同时，水力压裂法还会带来其他方面的问题。在这个过程中，要用巨大的压力将沙、水和很多化学品的混合物压入地下。这会破坏页岩层的构造，致使储藏其间的气体被挤出地表。但是，开采天然气在 2005 年得到《安全饮用水法案》的法定豁免，现在水力压裂法的使用几乎没有任何监管。最近有三名国会议员就水力压裂法发表报告，报告指出，2005 年至 2009 年，石油和天然气公司使用的 2500 多种压裂材料含有 750 种化学物，有些成分对人体无害，如柠檬酸、速溶咖啡。但也一些含有剧毒，如苯和铅。

1)booming 2)controversial 3)fracking 4)estimated 5)cubic 6)tremendous
7)ferocious 8)considerably 9)ingredients 10)benzene

二、口语部分

1. 有用的句子

（1）压缩天然气（CNG）（在高压下存储的甲烷）可代替汽油、柴油和丙烷/液化石油气。天然气燃烧后产生的不良气体比上面提到的燃料更少。

（2）将天然气（主要由甲烷构成）压缩成比标准大气压下体积的百分之一还小的体积，称为 CNG。它被存储在压力为 20～25MPa 的压力容器中进行分销。CNG 压力容器通常为圆柱形或球形。

（3）CNG 用于传统的汽油内燃机汽车，这些汽车已被转换为双燃料汽车。

（4）压缩天然气汽车比传统汽油动力车辆需要更大的燃料存储空间。因为它是一种压缩气体，而不是液态汽油，每加仑汽油当量（燃烧一加仑汽油产生的能量）压缩天然气需占用更多的空间。

（5）因此，用于存储天然气的容器通常在一辆车的后备箱或使用 CNG 作燃料的货车车厢中占用额外的空间。

（6）CNG 专用汽车解决了这个问题，将 CNG 容器装在汽车底部，这样就不用占用后备箱。

（7）另一个选择是将其安装在汽车顶部（主要用于公共汽车），不过，这需要解决结构强度的问题。

（8）在发生泄漏时天然气比其他燃料安全。因为它比空气轻，所以被释放后会很快消散。

2. 对话

A：李先生，我能耽误您几分钟时间吗？我有几个关于 CNG 的问题。

B：请问吧。

A：作为一种替代燃料，CNG 有什么优缺点呢？

B：每年有 120000 辆天然气燃料汽车行驶在美国的高速公路上。压缩天然气相对于普通汽油有很多优势。相比传统的柴油或汽油汽车，天然气汽车废气排放较低。天然气汽车车主指出，他们的车使用寿命更长，通常不需要太多的维护。天然气汽车的燃料经济性、加速性能和行驶速度通常不差于传统汽车。

A：天然气动力汽车安全吗？

B：据美国能源部所说，总的来说，天然气汽车和汽油或柴油燃料的汽车一样安全。

天然气汽车燃料箱设计为可承受极端的温度和剧烈撞击。在发生撞车时，压缩天然气会蒸发到空气中去，而不是像汽油或柴油那样聚集在车辆下面。这可能是一个安全特性，因为蒸发的天然气不太可能在从容器泄漏后被点燃。

  A：选择驾驶天然气汽车有什么局限性吗？

  B：如果你想要买一个天然气汽车，在美国选择面很窄。福特、通用汽车和道奇公羊计划将来某时引入双燃料天然气汽车，这意味着汽车可以同时由天然气和普通汽油提供动力。不过本田已经有了一款在售 NGV 车型。但 NGV 比传统汽车和卡车更贵，事实上，本田 NGV 为这一项功能就得额外多花 5000 美元。专家指出，NGV 行驶里程有限，比汽油动力的车辆要少 130 英里。

  A：还有其他不利因素吗？

  B：美国全国约有 1000 个天然气加气站，然而，可供公众使用的只有大约 540 个。此外，这些加气站建在城市中心，而不是在高速公路上。家用天然气加气设备花费在 2000 美元和 5000 美元之间，而且需要一个晚上才能给你的车加足燃料。一旦天然气储备枯竭，国家不得不到国外寻找其他燃料来源。

三、阅读部分

压缩天然气在亚洲的发展

  在新加坡，越来越多的公共交通工具（比如公共汽车、出租车以及货车）在使用压缩天然气。

  缅甸的交通运输部 2005 年通过一项法律，要求将所有公共交通工具——汽车、卡车和出租车改装为以压缩天然气为燃料。政府允许几家私人公司来对现有的柴油和汽油汽车进行改造，也开始进口压缩天然气为动力的公交车和出租车。

  在中国，像中能集团这样的公司在积极扩大压缩天然气加气站在中国内陆中型城市中的覆盖面，在内陆地区至少有两个天然气管道在运行。

  在巴基斯坦，卡拉奇政府在 2004 年最高法院的命令下强制要求所有城市公交车和机动三轮车上使用压缩天然气，以减少空气污染。

  在印度，德里运输公司经营压缩天然气动力汽车车队。

  在巴基斯坦，由于天然气短缺，且已对制造业造成负面影响，2012 年联邦政府宣布计划在三年时间内逐步停止使用 CNG。

  伊朗有世界上最大的 CNG 汽车车队和 CNG 分销网络。有 1800 个 CNG 加气站，共有 10352 个天然气喷嘴。在伊朗使用压缩天然气作动力的车大约有 260 万辆。

四、翻译部分

  (1) One problem remains, that is, setting up the network of fueling stations that can support a growing fleet of NGVs.

（2）Italy currently has the largest number of CNG vehicles in Europe and is the 4th country in the world for number of CNG-powered vehicles in circulation.

（3）We will discuss the technological problems in CNG fueling station, including metering, compression, dehydration, gas storage and high pressure pipeline.

（4）But in the course of using natural gas as fuel, CNG/gasoline bi-fuel engine may appear some problems such as the engine wear and tear with corrosion, the decrease of engine oil service life, etc.

（5）Compressed natural gas is often confused with LNG (liquefied natural gas). While both are stored forms of natural gas, the key difference is that CNG is gas that is stored at high pressure, while LNG is stored at very low temperature, becoming liquid in the process.

（6）与储存液化天然气相比，压缩天然气的生产和储存成本较低，因为它不需要昂贵的冷却过程和低温容器。存储和汽油或汽油同质量的 CNG 需要更大的容量并使用非常高的压力（20.5 到 27.5MPa，或 205 到 275 巴）。

（7）液化天然气通常用于通过船舶、火车或管道进行远距离运输，然后转化为压缩天然气分配给最终用户。

（8）CNG 正在被实验性地储存到内部压力较低的——约为 35 巴（3.5MPa，气体在天然气管道中的压力）的吸附式天然气储罐中，使用类似海绵的材料（如活性炭、金属有机多孔骨架）。

（9）ANG 存储时的能量密度与 CNG 相似或更大。这意味着车辆可以直接通过天然气管网加气，而不需要再压缩。燃料储罐可以做得更小，并可以使用更轻、更弱的材料。

（10）压缩天然气有时与氢气混合 (HCNG)，可增加燃料的 H/C 比值 (热容比)，使火焰传播速度比 CNG 快大约 8 倍。

References

[1] 杨秀华 . 油气储运专业英语 . 北京：石油工业出版社，2014.

[2] 段常贵 . 燃气输配 .4 版 . 北京：中国建筑工业出版社，2011.

[3] 唐秀岐 . 燃气输配与运营管理 . 北京：石油工业出版社，2012.

[4] 谭洪艳 . 燃气输配工程 . 北京：冶金工业出版社，2009.

[5] 花景新 . 燃气场站安全管理 . 北京：化学工业出版社，2013.

[6] 李庆林，徐鬻 . 城镇燃气管道安全运行与维护 . 北京：机械工业出版社，2014.

[7] 白世武 . 城市燃气实用手册 . 北京：石油工业出版社，2008.

[8] 袁宗明 . 城市配气 . 北京：石油工业出版社，2004.

[9] 城镇燃气设计规范：GB 50028—2006.

[10] 城市热力管网设计规范：CJJ 34—2010.

[11] 城镇燃气工程基本术语标准：GB/T 50680—2012.

[12] 中国城市燃气协会 . 城镇燃气及燃气具标准规范汇编 . 北京：中国建筑工业出版社，2013.

[13] 向阳 . 建筑类专业英语：第 2 册：暖通与燃气 . 北京：中国建筑工业出版社，1997.

Appendix Listening Materials

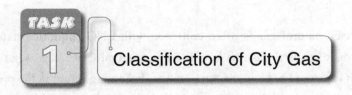

City gas or urban gas refers to any type of fuel gas which meets the normative gas quality requirements, and is supplied to meet the needs of residential, commercial and industrial users, generally including natural gas, liquefied petroleum gas (LPG), artificial gas and biogas.

Natural gas is a gaseous fossil fuel found in oil fields, natural gas fields and coal beds. It is the result of decay of animal remains and plant remains that have occurred over millions of years. The primary component of natural gas is methane, and it also contains gaseous hydrocarbons such as ethane, propane and butane, as well as other non-hydrocarbon gases.

LPG, one of the main sources of city gas, is obtained as a by-product during the process of exploitation of natural gas and petroleum or petroleum refining. The main components of LPG are propane, propene, butane and butene.

Artificial gas refers to the combustible gases made from some solid or liquid fuel and produced through all kinds of hot working. Based on different raw materials and processing methods, it can be divided into coal gas and oil gas.

Various organic substances, such as proteins, cellulose, fat, starch, etc., ferment in the absence of air, and produce a type of combustible gas under the action of microorganisms, which is called marsh gas (biogas), and can be divided into two categories: natural and artificial.

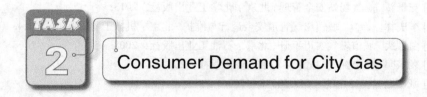

With the development of gas industry, especially the significant exploitation and utilization of natural gas, gas has become an important pillar of energy supply. According to the characteristics of users' using gas, city gas users generally include the following types:

Residential Users

Residential users refer to the household gas users who use gas as the fuel for cooking and preparing hot water. They are one of the basic users of city gas supply, and require continuous and stable gas supply.

Commercial Users

Commercial users are another basic user of city gas supply. They refer to the gas users who use gas for cooking or preparing hot water in commercial facilities or public buildings, including staff canteens, catering industry, kindergartens, nurseries, hospitals, hotels, barber shops, baths, laundries, supermarkets, offices, research institutes, universities, secondary schools, and primary schools etc. In schools and research institutes, gas is used in laboratories, as well as for cooking, hot water, and shower.

Industrial Production Users

Industrial production users are those who take gas as the fuel for industrial production. The gas consumption of these users is usually large and regular.

Heating, Ventilation and Air Conditioning Users

They are the users taking gas as the fuel for heating and cooling, and they are seasonal load to gas supply, so there must be effective measures to balance the uneven gas requirement in different seasons.

Gas Vehicle Users

These users are those who use gas as the automobile power. Developing gas vehicles is one of the effective measures to reduce urban air pollution. In addition, gas has an obvious advantage over petrol in terms of price.

Other Users

Other users mainly include two parts: one is the amount of pipeline leakage due to external damage, natural corrosion, improper use, production venting and other factors; the other part is the volume exceeding the original gas volume calculated, owing to new developments unforeseen.

Thermal Power Plants

When power plants use gas as the fuel for peaking and generation, urban gas load should also include the gas consumption in power stations. Converting clean-burning natural gas to electrical energy with zero discharge of pollutants is a major development direction of natural gas applications.

Data shows that natural gas has begun to be applied and developed in greenhouses planting flowers and vegetables, grain drying and storage, deep processing of agricultural products, biotechnology, pharmaceuticals, pesticides, and gas fuel cell lights, etc., which will lead to continuous expansion and more detailed classification of urban gas users.

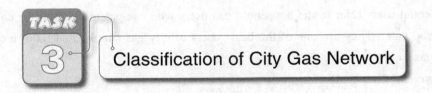

TASK 3 Classification of City Gas Network

Classification of Gas Pipeline

First of all, the city gas pipeline according to the gas transmission pressure can be divided into high pressure A and B gas pipeline, sub-high pressure A and B gas pipeline, medium pressure A and B gas pipeline, and low pressure gas pipeline. Secondly, according to the use the city gas pipeline is divided into gas transmission pipeline, gas distribution pipeline, building service pipeline, indoor gas pipeline and gas pipeline of industrial enterprises. Thirdly, the city gas pipeline is divided into the underground gas pipeline and overhead gas pipeline according to the laying mode. Fourthly, the city gas pipeline, according to the shape of pipe network, is divided into circular pipeline network, branched pipeline network and circular and branched pipeline network.

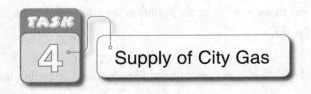

TASK 4 Supply of City Gas

Storage and Transportation of City Gas

Generally speaking, the commonly used storage methods of natural gas include gas storage tank, gas underground reservoir, gas pipeline and tube bundle, liquefied natural gas storage. Gas storage tanks are divided into low pressure gas storage tanks and high pressure gas storage tanks. The underground storage of natural gas usually includes the depleted oil and gas fields storage, aquifer porous formation storage, salt layer storage and cave storage, etc. At present liquefied natural gas storage generally adopts the method of low temperature and atmospheric pressure storage. Other storage methods of natural gas include the storage in low temperature liquefied petroleum gas and solid-state storage etc. Transportation modes of city gas include pipeline transportation, rail tanker transportation, road tanker transportation and waterway transportation.

Gas Network Peak Shaving

Supply and Demand Balance of City Gas

City gas consumption varies with time, the monthly, daily and hourly are not the same, but the gas sendout of gas source generally changes little, especially long-distance gas transmission pipeline. Therefore, unbalance between gas sendout and gas consumption often occurs. In order to ensure supply according to the requirements of customers, we must solve the unbalance problem of gas sendout and gas consumption. To solve the problem of the monthly (quarterly), daily or hourly uneven gas consumption we can adopt different storage methods, such as underground storage, gas storage tank, liquid-state storage and gas transmission pipeline end storage etc. Underground storage is mainly used to overcome the seasonal uneven gas consumption, and gas storage tank, liquid-state storage and gas transmission pipeline end storage are mainly used to solve the unbalance problem of hourly gas consumption.

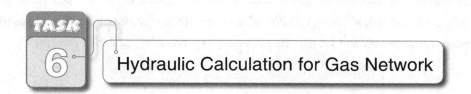

Hydraulic Calculation for Gas Network

The Computer Application in Hydraulic Calculation

Hydraulic calculation for city gas network is one of the main jobs in city gas design. The common methods include loop analysis method and node flow method (also known as the hydraulic calculation method). The first method only applies to small branched networks, the second method has the characteristic of gaining true solution, through iteration, under the circumstance of ignorance of pipe section flow, which applies to all kinds of large-scale and complex gas network, but it has so large computational effort that hand computation is difficult, and therefore, it is usually operated by computer. Based on AutoCAD, the reading of section node coordinates of gas network is realized by VB secondary development technology and hydraulic calculation drawing of gas network is automatically generated, and the hydraulic calculation software for city gas network can be programmed by VC++ visual programming language.

Natural Gas Storage and Distribution Station Process

The processes of two-stage pressure regulation, gas storage of high-pressure and gas supply of sub-high pressure are adopted in the natural gas storage and distribution station. Pressure of high-pressure main pipe of natural gas, which flows from gas distribution station to factory storage and distribution station, is 0.5 ~ 1.5MPa, this pressure turns to be 0.3MPa after it is regulated in the storage and distribution station, which flows in the network of gas supply of sub-high pressure for industrial and civil use. When the storage and distribution station is low-peak load, natural gas getting in the station by measuring and regulating is supplied to pipe network of sub-high pressure directly, and at the same time, it is made to take the aeration to gas storage tank. When gas pressure getting in the station is lower than that of gas storage tank, it is needed to use the compressor to pressure up to that of gas storage tank, and then the gas is made to take the aeration to gas storage tank. In general, pressure of gas storage tank is 0.8 ~ 1.5MPa. When users are in high-peak load, natural gas must be let out from gas storage tank and get into the gas supply pipeline after regulating pressure.

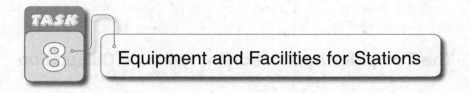

Gas Storage Facilities

Gas demand of a city is not constant and varies with time. But gas source supply is almost constant and does not change with time. To solve the confliction of constant supply and inconstant demand, city gas distribution system is equipped with gas storage station. Gas storage station stores gas during low demand and releases gas at peak demand. The following are several commonly used storage methods. At low gas demand season, gas is pumped into underground space with proper geological structure. At high gas demand season, gas is withdrawn from the underground space. The proper geological structures are usually exhausted oil or gas fields, underground structure containing water and porosities, salty ore beds and caves, etc.

The most economical method is to store city gas in exhausted oil or gas fields. Underground storage reservoirs can store huge volume of gas with small investment and operation costs. This method saves thousand tons of steel materials. It is usually used for seasonal peak-shaving and partial daily peak-shaving. It is an ideal storage method for city gas. The difficult part is how to find an economical and suitable geological structure near the city to store gas. The capacity of the first underground storage reservoir in China is 10 million cubic meters. Large scale underground storage reservoirs have been under construction since the beginning of Western to Eastern Project. The volume of liquefied natural gas is much smaller, 1/600 volume of natural gas of the same mass. It can be stored in a heat insulated holder. Liquefied natural gas will be gasified at gas demanding peak hours and the gas will be supplied to users.

Now LNG has become an international commodity because it is easy to be transported and traded between countries. Recently, global LNG production and trade are getting active. LNG is booming in oil and gas industry.

Gas storage by high pressure pipelines and end section of long transmission pipelines is an effective way of hourly peak-shaving. Groups of high pressure pipelines are buried underground and the natural gas inside them is compressed to high pressure to be stored. End section of long transmission pipelines can store certain amount of natural gas during night time when natural gas demand is low and release natural gas at daytime when natural gas demand is high.

The most commonly used gas storage method in China is to store gas with low-pressure or high-pressure gasholders to balance daily or hourly gas demand fluctuations. Compared with other storage methods, gasholder storage consumes more metal materials and costs more. But when other gas storage methods are not available, gasholder storage is the only way to store gas.

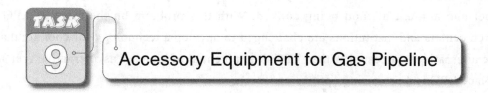

Water Drainers

Water drainers are used to discharge the condensed water continuously in the pipeline network in the case of normal water seal height on which gas can't be discharged. Therefore, water drainers are necessary safety affiliated facilities. Usually overhead pipeline drainers can be built into vertical and horizontal drainers. In order to keep the effective height of drainers and easy operation and maintenance, there are usually single water seal and double water seal forms.

Gate valves should be installed for the drainers of gas pipelines, and be vertical connection with a collection funnel drain, and a drainer is feasible, but that will increase partial burden for the pipeline when its foundation sinks. Drainers can be located outdoors, but anti-freezing measures must be taken in cold areas. When they are located indoors, there should be a good natural ventilation.

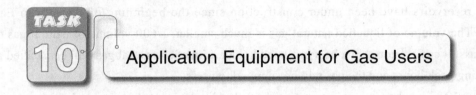

TASK 10: Application Equipment for Gas Users

Gas Cookers

Liquefied petroleum gas (LPG) is one of the conventional sources of fuel for cookstoves in the Philippines. The use of LPG as source of fuel is common both in the urban and in the rural areas, particularly in places where its supply is readily accessible. The main reasons why LPG is widely adopted for household use are: it is convenient to operate, easy to control, and clean to use because of the blue flame emitted during cooking. However, because of the continued increase in the price of oil in the world market, the price of LPG fuel has gone up tremendously and is continuously increasing at a fast rate. At present, an 11kg LPG, which is commonly used by common households for cooking, costs as high as P540 per tank in urban areas or even higher in some places in rural areas. For a typical household, having four children, one LPG tank can be consumed within 20 to 30 days only depending on the number and amount of food being cooked. With this problem on the price of LPG fuel, research centers and institutions are challenged to develop a technology for cooking that will utilize alternative sources other than LPG. The potential of biomass as alternative fuel source to replace LPG is a promising option.

Appendix Listening Materials

Task 11: Gas Pipeline Construction

A Pipeline Deal to Exploit America's Fast-Changing Energy Landscape

Nodding donkeys, offshore platforms refineries and filling stations are the bits of the oil industry you can see. A vast and largely invisible network of underground pipes joins them all together. It is worth a lot, which is why Energy Transfer Partners (ETP) said it would pay $5.3 billion for Sunoco on April 30th. It hopes to pull together two networks and shift more of America's booming oil and gas output.

ETP is ambitious. Last year its parent company, Energy Transfer Equity, agreed to buy Southern Union and its gas-pipeline network for $5.7 billion. The latest deal will make ETP the country's second-biggest pipeline firm, behind Kinder Morgan, after the latter's merger with El Paso is concluded later this year.

Sunoco comes with storage facilities, 4,900 filling stations and the remains of a refining business that it is trying to spin off in a joint venture with Carlyle, a private-equity firm. But the pipelines are the main attraction. ETP currently operates 17,500 miles (28,160km) of the arteries that transport gas and natural-gas liquids such as propane and butane. Adding Sunoco's 6000 miles, built to carry crude oil and refined products, will reduce ETP's reliance on gas. After the deal, 30% of its revenues will come from oil.

Task 12: Quality Supervision and Inspection for Gas Pipeline

The Battle over Gas Prices in Europe

NO one likes getting a gas bill. But Europe's biggest utilities are especially upset over the sums they must pay gas producers, in particular Russia's state-backed giant, Gazprom. Some are trying to cut those costs, but with little in the way of leverage over producers their chances of success look slender. Gas prices in continental Europe are mainly set by a decades-old system of long-term contracts, linked to the price of oil. But in relatively liberalised Britain gas is largely traded at spot prices set by current supply and demand. This handed an advantage to some smaller European utilities with interconnections to the British spot market when, in 2008, gas prices suddenly fell. Their bigger rivals meanwhile suffered. They were saddled with "take or pay" contracts that obliged them to buy fixed quantities of gas far above what they could sell and at prices way above those on the spot market. The gap between spot and contract prices has not gone away. German firms, which are especially hostage to Russian pipelines, are at a big disadvantage. Yet both Russia and Norway, which supply almost half of Europe's gas, have shown some flexibility towards their complaining customers. By the end of 2009 European gas buyers were begging for relief, with oil-indexed gas then 50% pricier than spot gas. In response, Norway's Statoil allowed an element of spot pricing (around 25% of oil-indexed contracts) for three years. Gazprom responded with a less generous 15% ~ 20% allowance.

TASK 13 Station Management

Operation and Management of City Gate Station

City gate station is the terminal gas distribution station of long-distance gas transmission pipeline, generally located at the end of long-distance gas transmission pipeline and the beginning of city gas pipeline network. It is the important facility of city gas transmission and distribution system and also the starting point and total hub of city gas transmission and distribution system. The staff of city gate station should strictly implement the regulations of the security work and fire safety work focused on the vital parts. The staff of city gate station should strictly implement the routing inspection, duty and shift system. At the same time the staff must be familiar with the various processes and the testing, installation, operation, maintenance, repair of the equipment used in the processes, and must operate in accordance with the equipment operating procedures, maintenance and overhaul procedures and safety technology regulations. City gate station should be equipped with the full-time safety management personnel for operation, maintenance and repair, and should establish the corresponding safety target responsibility system, safety production management system, safety production responsibility system of post operation and emergency repair plan. The operation and management of city gate station should be consistent with the current national mandatory standards.

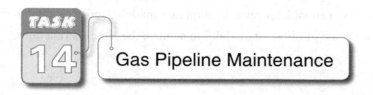

Gas Pipeline Maintenance

Routine Maintenance of City Gas Pipeline

Routine maintenance of city gas pipeline is the assurance of the normal operation of the pipeline. The main work of routine maintenance is generally divided into two categories: one is to maintain the pipeline itself, namely the routing inspection (including leak detection, leakage treatment), the valve chamber maintenance, pipeline position detection; the other is to deal with the influence of external factors on the pipeline, for example the treatment of the third-party slight destruction and damage, illegal buildings. On the city gas pipeline (including polyethylene pipeline), the staff regularly carries on the routing inspection, leak inspection and pipeline inspection, which is an important work of the routine maintenance of city gas pipeline network. A decision about the frequency of the routing inspection is made according to the region classification, pipe material and operating pressure. Meanwhile, the valves, condensate drainages, expansion joints and other accessories installed on the city gas pipeline should also be under strict routine maintenance.

Safety Management for Gas Stations

Type and Function of Gas Stations

City gas transmission and distribution system is the whole system which starts from city gate station receiving gas supply of the long distance transmission pipeline to the user equipment, and it includes city gate station, gas storage and distribution station, and so on. City gas stations play a very important role in city gas transmission and distribution system, and generally have the functions of storage, vaporization, pressure regulating, metering and odorization, etc. According to their functions, city gas stations include city gate station, regulator station, gas storage and distribution station, vaporizing station, gas mixing station, gas fuelling station, and so on. The technological process and equipment and facilities of city gas stations are also different with their functions different. We can look on each function as a module, and according to the actual needs, combination can be made of different modules to achieve the required functions.

TASK 16 — Safety Management for Gas Pipeline

Significance of Safety Management for Gas Pipeline Network

The pressure pipeline in city gas transmission and distribution system is a kind of special equipment which is widely used in production and daily life and which may cause the higher danger of explosion or poisoning accident. The characteristics of gas pipeline as a kind of pressure pipelines are that gas pipeline is mostly buried underground, and construction, inspection and maintenance of gas pipeline are difficult through densely populated areas, limited by geographical conditions and affected by external causes, it is easy for gas pipeline to cause fire, explosion or poisoning accident to result in the larger social impact and damage. Therefore, the existing security problems of gas pipeline are analyzed and effective preventive measures are put forward to ensure the secure and stable operation of gas pipeline network, which will have very important practical significance.

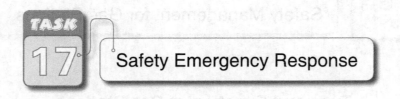

TASK 17 — Safety Emergency Response

Troubleshooting of Gas Pipeline

In the operation of gas pipeline, the most likely problem is the leakage and pipeline blockage, and timely discovering leakage is the premise of the prevention and treatment, and is one of the main tasks of operation management for gas pipeline network. So in the management of gas pipeline, we should focus on catching two problems of the leakage and blockage of gas pipeline, as well as the troubleshooting caused by the leakage and blockage of gas pipeline. The staff should carry out the on-line inspection of pressure pipeline according to the relevant rules and requirements, and find problems in time and eliminate the hidden dangers to ensure the smooth and safe operation of gas pipeline. At present, the commonly used methods of leak detection in the world include the visual leak detection method, sound leak detection method, the sense of smell leak detection method and tracer leak detection method.

TASK 18 Liquefied Natural Gas

Falling Oil Prices Affect Nations Differently

World oil prices have dropped by about 30 percent since June. That has helped many economies around the world. When people spend less on fuel, they can spend more on other goods. But lower oil prices also mean reduced profits for national oil companies. Privately-owned oil producers and companies that provide services to the oil industry are also affected. Increased American oil production is one reason for the drop in world oil prices. Technology developed in the U.S. has made that increase possible. Hydraulic fracturing, or fracking, is a process of pumping water and chemicals into the ground to force out natural gas and oil. It is often used to recover these fossil fuels from rock called shale.

Jim Krane is an energy expert at the James Baker Institute for Public Policy at Rice University. He says oil produced from shale has changed the market.

Slowing economies in China and Europe have also reduced the demand for oil. In the past, the Organization of Petroleum Exporting Countries would react to decreasing demand by cutting production. The twelve member states often cooperate to influence oil prices around the world. However, Jim Krane says members want to defend their share of the market and are unwilling to make production cuts. "They are willing to see prices drop for a little while to see if they could sweat some of these higher-cost producers out of the market." However, this could affect small oil producing companies in the U.S. These companies feel the effects of falling prices more sharply on their profit margin. "If you are a small company thinking about investing in putting in some new oil production, well, you might think twice if prices are down at $75 (a barrel)" U.S. companies are not alone in feeling pressure from falling prices. Oil makes up 83 percent of

Nigeria's exports and 70 percent of the nation's economy. Finance minister Ngozi Okonjo-Iweala recently announced measures the government would take to increase income.

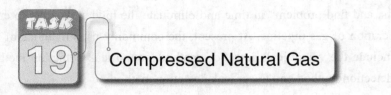

TASK 19 Compressed Natural Gas

The Need to Be Seen to Be Clean

Natural-gas production is booming, but its green image is in question.

There are the signs of America's natural-gas boom. Thanks to new drilling technology, and in particular a controversial process called hydraulic fracturing or "fracking," the size of the proven reserves is growing. At the end of 2009 the United States had estimated reserves of 283.9 trillion cubic feet (8 trillion cubic meters) of natural gas, up 11% from the year before. In 2010 the country produced 22.6 trillion cubic feet of natural gas, up from 18.9 trillion cubic feet in 2005. The price at the wellhead has dropped from $7.33 per thousand cubic feet to $4.16 during the same period.

But some question whether natural gas is really as green as all that. For one thing, fracking uses a tremendous amount of water, a severely undervalued resource inland. On the other hand, the process gives off methane, a potent heat-trapper. And fracking raises other concerns. In the process, a mix of sand, water and chemicals is pumped deep underground at ferocious pressure. That breaks up considerably formations, releasing the gas trapped inside so it can be pumped to the surface. But fracking is almost entirely unregulated, because of a 2005 statutory exemption from the Safe Drinking Water Act. Three members of Congress recently released a report on fracking, saying that oil and gas companies used more than 2,500 fracking products containing 750 chemicals between 2005 and 2009. Some ingredients were benign, such as citric acid and instant coffee. Others, though, were extremely toxic, such as benzene and lead.